A World History of Mobility

Explorations in Mobility

Founding Editor
Gijs Mom, Eindhoven University of Technology

General Editors:
Cotten Seiler, Dickinson College
Mimi Sheller, Drexel University

The study of mobility opens up new transnational and interdisciplinary approaches to fields including transport, tourism, migration, communication, media, technology, and environmental studies. The works in this series rethink our common assumptions and ideas about the mobility of people, things, ideas, and cultures from a broadly understood humanities perspective. The series welcomes projects of a historical or contemporary nature and encourages postcolonial, non-Western, and critical perspectives.

Volume 8
A World History of Mobility: An Essay on Road Cultures and Beyond
Gijs Mom

Volume 7
If Cars Could Walk: Postsocialist Streets in Transformation
Edited by Ger Duijzings and Tauri Tuvikene

Volume 6
Transnational Railway Cultures: Trains in Music, Literature, Film, and Visual Art
Edited by Benjamin Fraser and Steven Spalding

Volume 5
Iron Landscapes: National Space and the Railways in Inter-War Czechoslovakia
Felix Jeschke

Volume 4
A U-Turn to the Future: Sustainable Urban Mobility since 1850
Edited by Martin Emanuel, Frank Schipper, and Ruth Oldenziel

Volume 3
Driving Modernity: Technology, Experts, Politics, and Fascist Motorways, 1922–1943
Massimo Moraglio

Volume 2
The Devil's Wheels: Men and Motorcycling in the Weimar Republic
Sasha Disko

Volume 1
Atlantic Automobilism: The Emergence and Persistence of the Car, 1895–1940
Gijs Mom

A World History of Mobility

An Essay on Road Cultures and Beyond

Gijs Mom

berghahn
NEW YORK • OXFORD
www.berghahnbooks.com

First published in 2025 by
Berghahn Books
www.berghahnbooks.com

Library of Congress Cataloging-in-Publication Data

A C.I.P. cataloging record is available from the Library of Congress
Library of Congress Cataloging in Publication Control Number: 2025937177

British Library Cataloguing in Publication Data

A catalogue record for this book is available from the British Library

EU GPSR Authorized Representative

LOGOS EUROPE, 9 rue Nicolas Poussin, 17000, LA ROCHELLE, France
Email: Contact@logoseurope.eu

ISBN 978-1-83695-091-2 hardback
ISBN 978-1-83695-092-9 epub
ISBN 978-1-83695-093-6 web pdf

https://doi.org/10.3167/9781836950912

She walked until she reached the big crossroads and waited at the kerb while the traffic roared from one direction and then the next. Twice she stepped into the road and drew back again. To get to the other side of the street without being hit by a car was like walking out in the monsoon and hoping to dodge the raindrops. A space opened up before her. God is great, said Nazneen under her breath. She ran.

A horn blared like an ancient muezzin, ululating painfully, stretching his vocal cords to the limit. She stopped as the car swerved. Another car skidded to a halt in front of her and the driver got out and began to shout. She ran again and turned into a side street, then off again to the right onto Brick Lane.

—Monica Ali, *Brick Lane*

Contents

Preface

The idea for this book came to me after a discussion with a Chinese PhD student at my university in which I had advised her to look at my trilogy on world mobility history. She could not find what she wanted to know, despite the very detailed index. In the Introduction I will explain how I conceived the book's content, but the wish to create an entrance into the trilogy for novices in New Mobility Studies (NMS) formed the trigger.

The publication of this book, less than half a decade after the completion of the manuscript of the third volume, takes place in turbulent times in both real-world mobility and New Mobility Studies. At the closing of my world history project, I wish to acknowledge the people that, knowingly or unknowingly, enabled me to develop its content well beyond the nearly three thousand pages of text and an even larger number of sources that made up the trilogy, plus my dissertation about the history of the electric vehicle. Returning to the annual conferences of the International Association for the History of Transport, Traffic and Mobility (T2M) in Seoul (2023) and Leipzig (2024, both co-organized by Lancaster University, and both attended by mobility historians and social scientists), I found a new generation of mobility scholars as well as a group of literary scholars specializing in mobility, whose presentations taught me about new approaches, new geographic areas of study, and even new vehicles, from cycling to informal transport, from electric trolleybuses to marginalized groups, and from Asia to the 'Global East.' An invitation to participate in a conference at the University of Konstanz in October 2024 deepened my understanding of the (American) road narrative. The presentations at these events, many of them not yet developed into peer-reviewed texts, enabled many additions and corrections to my project.

I am grateful and proud that the NMS field is so vibrant and still expanding into geographies and scholarship beyond the West. I am also grateful to the young scholars I met at my own university in Eindhoven, such as Frauke Behrendt and Jonas van der Straeten, who, together with Ruth Oldenziel, are busy expanding and deepening the field of cycling studies toward a global project. On the more practical side I received invaluable support from the international library loan team Peter Smits and Marly van Gog at Eindhoven University of Technology. I am also grateful to the Berghahn people, from Marion Berghahn for her trust in this project, to Amanda Horn, Emma Wallace, and Keara Hagerty for their guidance during the publication process. I also thank the two anonymous reviewers for their helpful comments (which made the manuscript grow by 10 percent) and Harry Eagles for copy-editing the text. Texas van Leeuwenstein helped me compile the index.

Oh, and while researching and writing I did not use AI or ChatGPT.

Órgiva, 1 November 2024

INTRODUCTION

Making World Mobility History

This book is intended to be, and has been written as, a (somewhat opinionated and personalized) *essay*, rather than a formal scholarly historical study, in order to answer a seemingly simple question: Why the car (and why not)? While my answers to the first part of the question will necessarily be historical, the second part of the question suggests a look into the future, based on the answers to the first part. To reach both aims it was clearly not enough to rewrite my monograph dealing with the history of the electric vehicle[1] and the trilogy on global automobilism,[2] written during the past quarter century, into a shortened version in which the argumentation takes center stage (and referencing in the footnotes has been limited to justifying direct quotations, and new research).[3] Such a shortened version is presented here, but in addition it was necessary to expand the historical evidence into three directions.

First of all, whereas the explanation of the decline of the early electric car hinged on the masculine character of the hegemonic car culture, the automobilism trilogy, which began with the same assumption, shifted in the course of its three volumes toward an acknowledgment of another trend, less well covered by mobility history: the status character of the car as commodity, a trait that emphasized ownership rather than use, and became more pronounced as car culture became global, a trend I sometimes called 'feminized,' to contrast it with the dominant masculinized user culture of the car as 'adventure machine.' It appeared appropriate, therefore, that some extra research would be done into the role of a minority group of motorists (mostly women, but also some men) in the production of another, less aggressive and violent adventure, expressed in the early 'gymkhanas' and flower contests and revived half a century later in a youth culture of car modifying and 'lowriding.' The writing of the third volume of the trilogy brought the insight that without reconstructing this 'feminine' and alternative 'masculine' tradition of the automotive culture, such new subcultures could not be satisfactorily explained. Through exploring, in a more pronounced way, the role of women in early automobilism, and some trends of nonhegemonic masculinity in the West as well as redefinitions of the masculine self in Asia in the post-World War II phase, the 'consumerist' tendencies in global automobilism could be given more relief and the fledgling contours of a perhaps different car culture in the Global South could more convincingly (I hope) be identified. In other words, this essay is not so much a gendered history of automobilism as a more gender-equal base for such a history.

For the same reasons, some extra attention is given to the bicycle culture at the turn of the twentieth century, prefiguring in many ways the 'soft' elements in the subsequent culture of automobilism, elements that did not disappear with the emergence and persistence of the car. This is all the more urgent given the recent revival of interest in cycling as an alternative to and a critique of the car. Likewise, the motorcycle had to gain somewhat more attention because of its masculine reputation. This in its turn led to somewhat more emphasis on the class character of mobility cultures, not only to give (motor)cycle culture a clearer class profile (especially the role of the working and lower-middle class in this culture), but also to make the role of the global new middle classes more pronounced in the emergence of a worldwide car culture. Further, I used the opportunity of this essay to bring the narrative of the trilogy, which ended around 2015, up to date by including the last decade or so.

Additional research was also needed to make two countries a bit more prominent than in the original studies, Japan and the Soviet Union/Russia. Within mobility history, not much is known about the distinctive trajectory of motorization in Japan and its intermediary role as an Asian imperial power and, later, as a mediator between the West and the East in the neoliberalization of automotive Asia. Likewise, the Soviet Union has been under-researched because of the obsession with 'scarcity' among its mainstream 'area students,' neglecting the futile efforts in the SU and Eastern European socialist states toward an alternative motorization pattern, a pattern that again became more clear when writing the third volume. Together with very recent new research about Tanzania, I am now able to construct at least a patchwork of a possible 'socialist' car-cultural narrative, which can be connected to the efforts to construct alternative car futures, or futures alternative to the car, as discussed in the fully new Chapter 5. A visit to the 2024 T2M conference in Leipzig opened my eyes to the efforts to develop a 'socialist city planning' (as both a planning of socialist cities and a socialist planning of cities) including the proliferation of dozens of trolleybus networks, not only in the Soviet Union, but also in several Eastern European countries.[4]

Second, as the essayistic, succinct character of this book lends itself to reading in an educational context, the relevant theoretical tools had to be introduced more explicitly, in an abbreviated introductory form suitable for a classroom environment, located where they need to be understood for the clarification of the interpretation of particular historical evidence. They are (opinionated) vignettes rather than theoretical treatises, signaling the opportunity for further study. Also, for this purpose the somewhat enumerative presentation of hundreds of belletristic utterances had to be considerably diminished and its argumentative treatment tightened. Furthermore, the structure of the entire essay has been construed such that it would fit in a course setting through its seven independently readable texts (Introduction, Conclusions, and five chapters in between), whereas an annotated 'canon' should enable the novice in the history of mobility to become familiar quickly with the field's assumed common denominators, a necessity that will be made clear when dealing with the recent 'mobility turn' in Chapters 4 and 5.

Thus, a brief introduction to the social construction of technology and some other concepts from the history of technology and innovation theory (such as the concept of 'script' and the 'ironic car'), as well as a brief reference to gender and masculinity scholarship, are to be found in the first chapter, whereas the second chapter includes some paragraphs on the so-called 'toy-to-tool' thesis (the questionable thesis that the popularity of the car is due to its becoming a utilitarian tool) and the 'America as model' thesis (the no less questionable thesis that national histories of automobilism are following, somewhat later, the 'script' of the US in this respect), both relevant for understanding the *persistence* of car culture (rather than its emergence of the first period) in the West. Similarly, 'actor network theory' is briefly introduced in Chapter 3, as it was developed in part through the analysis of the (failed) efforts to reintroduce the electric passenger car in France after World War II, whereas the emergence of car modification and street racing among the Atlantic youth invites some attention to Mikhail Bakhtin's concept of the 'carnivalesque.' Chapter 4 contains an introduction to 'mobility Orientalism,' the problematic, Eurocentric reflex to analyze mobility in the Global South as 'backward,' assumed to 'catch up' to the Western model as soon as prosperity reaches its population, or, worse still, to universalize the Western story into a 'global' one without much consideration for the fundamental differences between the two.

The third adjustment to my earlier monographs concerns a rethinking of the concept of 'world history,' a rethinking needed for a better understanding of the globalization of automobilism. World history or global history (terms I use interchangeably) has a long tradition as "a minority tendency," reaching back at least into the sixteenth century of the Mughal Empire.[5] Nowadays, three distinctive 'schools' can be identified:

world history as a history of everything that has happened in 'the world,' constructed as a patchwork of individual histories, such as the history of silk, or (elements of) mobility history, several of which are presented in the annotated bibliography at the end of this book; second, as a history of connections and the 'transfers' between different cultures and civilizations, for instance the spread of the 'idea' of automobilism as well as of its materialities such as bicycle technologies and bicycle paths, or cars and car-friendly roads; or third, as a 'perspective,' a way of finding the global in the local for instance, such as the emergence of so-called 'layered' urban mobilities in Africa in which the car adds a new dimension to what many observers experience as 'chaos.'[6] This book's world mobility history has elements of all three approaches.

Efforts to write non-Western world visions, such as the Latin American *dependencia* theory (with its radiance toward Africa and influencing Wallerstein's approach, see Chapter 3) and subaltern studies (trying to give people a voice who are not to be found as active agents in the sources, for instance Indian peasants, see Chapter 2), or the 'Orientalism school' (depicting visions from the South as 'deviation from a standard pattern,' see Chapter 4), enabled a more balanced, multi-perspectivist view.[7] This multi-perspectivism is especially relevant when world historians focus on the process of 'globalization' itself, a phenomenon that also has a history of several centuries, but that experienced a rather sudden intensification and acceleration from the 1970s in which (as this study claims) the proliferation of automobilism (especially through the construction of a worldwide network of car-friendly roads, described in Chapter 3) played a crucial role. It is a conceptual, teleological error (a *Denkfehler*, as Helmuth Berking posited)[8] to suggest that globalization is a deterritorializing process of unbounded flows; on the contrary, globalization is not a top-down process alone, it also settles down locally, from where it at the same time is modulated.

But the traces mobility has left in world history writing are remarkably thin, literally so: in Patrick Manning's standard handbook it encompassed exactly two paragraphs, mentioning animal traction, shipping, and the 'technology of empire' in a reference to Daniel Headrick's famous study (without even mentioning the dominating land-based technology of the railways).[9] Likewise, efforts to conceive a global history of technology as the scholarly roots of mobility history, for instance in Arnold Pacey's and Francesca Bray's thousand-years-long *Technology in World Civilization*, focused on invention, innovation, and engineering rather than on use. Hence, it is understandable that automobilism hardly plays a role in such studies, as its culture was, as a 'bricolage' of existing cultures and technologies developed during the previous century (see

Chapter 1), hardly anything new, despite claims to the contrary about its revolutionary character.[10] But even a very recent and very welcome effort to rewrite current globalization from the perspective of 'things' and their surrounding 'practices' appears to be deficient in presenting automobilism as a worldwide phenomenon: in Nigel Thrift, Adam Tickell, Steve Woolgar, and William Rupp's *Globalization in Practice* (2014) the chapter on road safety is reduced to a survey of localized speed limits in Anglophone countries (neglecting an entire bookshelf on transnational automotive safety technology, safety statistics, and WHO and UN safety epidemiology, to mention only some omitted aspects), whereas the chapter on 'mobility' lacks any reference to globalized automobilistic empirics.[11]

However attractive it seems to many novices to the history of mobility, even the 'mobility turn' of a quarter century ago (which we will introduce as an important *historical* factor in Chapter 4) does not seem to satisfy the obvious need for a historical approach, for instance in the combination of 'global networks' and 'global liquids' ('automobility,' migration, tourism), in sociologist John Urry's attractive but rather schematic model. A similar attraction resides in more historical (but perhaps somewhat teleological) approaches that use alternative integrative metaphors such as 'web' (by the American world historian William McNeill).[12] Recently, an edited volume on the *Global History of Techniques* provided a correction to such approaches, in which Arnauld Passalaqua, writing on "Mobility Techniques," observed that 'mobility studies' "have diverged further and further away from the technical dimensions of the systems that are being studied."[13] Mobility historian Kate McDonald has warned us in her analysis of Japanese 'railway imperialism' against "the centrality of movement to the traditional conception of modernity as an historical period," because it risks to "uncritically adopt . . . the terms of its subject . . . we should take care to treat the mobility of modernity as an argument, born of imperialism."[14] To heed her warning, positioning 'mobility' as such in 'world history' needs careful theorizing indeed. The field of 'mobility history' is not yet so far: this book gathers the main elements of such a history in perhaps still a somewhat kaleidoscopic narrative. I have no doubt that mobility history, in the coming years, will gain from an ever increasing inclusion of mobility cultures.

The current study, because of its 'imprisonment in English' (including its English translations from Russian and from Asian, African, and Latin American languages), runs the risk of what Margit Pernau saw as a new form of colonialism: I am one of the 'caste of global historians' not endowed with linguistic knowledge of extra-European societies.[15] This is especially problematic for a special kind of source mobilized to feel the pulse of global road motorization: belletristic utterances in high- and lowbrow genres, but also forms of popular culture such as movies, songs, and games. Although I am less interested in their representative contents than in their symbolic and especially structurally homological properties (a three-level style of analysis explained in the next chapter), the selection of these sources is not only biased by their language (Dutch, English, German, French, some Spanish), but also by the selection process of both the translating publishers and the mainstream sources for book reviews that I used, such as the *Times Literary Supplement* and the *New York Review of Books*. Many of these (what I will call) 'autopoetic' sources (that is: not necessarily books on car travel, but sources with 'autopoetic' scenes, explained in Chapter 1) are in themselves considered to be part of 'world literature,' even if some literary scholars oppose such epithets on the grounds of these sources' 'untranslatability.'[16] For similar reasons this study includes songs only (with some African and Chinese exceptions) as part of Western pop and rock culture and not what in Europe has been called 'world music' of the Global South; nor will we go far beyond Hollywood's and Western European film culture (and we will neglect Bollywood and Nollywood [nearly] totally, but will deal with some Chinese movies and TV series). Conversely, when in Chapter 4 gaming is introduced we will dedicate some space to Asian (especially Korean and Japanese) designs.

In the theoretical introductions to the monographs on which this book is largely based I have addressed several 'problems' that need rethinking or reprocessing for the purpose of the present essay. Some of those problems have been resolved by time, as it were. 'The Trouble with Travel Writing,' for instance, namely that it deals with the destination rather than including the 'way' to get there, seems to be addressed by new approaches inspired by the 'mobility turn' (see Chapter 5).[17] Also, considering the car as a 'commodity' (mobilized as a 'correction' to a more or less exclusively practice-theoretical approach to the car as an 'adventure machine'),[18] the acquisition and possession of which grants status and at the same time enables fetishism, is widely enough accepted as a possible angle of approach to automobilism to obviate separate attention in this chapter, although I still miss the contribution by media studies scholars helping mobility historians to analyze the car as a message, a medium rather than a transport vehicle.[19] The same argument of sufficiently wide acceptance applies, I contend, to my original efforts to convince (what I thought to be) a potentially skeptical readership of transport students of the explanatory power of 'narrative histories.'[20] I refer those readers who are interested in a deeper understanding of these aspects and arguments to the monographs mentioned earlier.

At first sight, the same also seems to apply to the use of literature as a historical source, considering the very recent revival of a humanistic interest among mobility students.[21] Yet there still exist some doubts, especially among literary theorists working in the tradition of the 'mobilities paradigm,' about the merits of literature (and other 'texts' such as movies or songs) as historical evidence, doubts that are fed by poststructuralist theorizing, most particularly by what human geographer Nigel Thrift has termed 'non-representational theories.'[22] In the first volume of my world mobility trilogy I have introduced, referring among others to the work of Thrift, several theories of 'social practices' (philosopher-geographer Theodore Schatzki's social action theory, the theory of affordance by psychologist James Gibson, and the 'hybrid subjectivity' of German cultural sociologist Andreas Reckwitz) as an entrance into narrativist histories of automobilism, theories that also inspire the current book. On this basis I have defended the use of literature as a historical source by pointing at the *performative* aspects of writing (and reading).[23] Hence my earlier emphasis on the 'affinity' of these practices with driving and passengering, as well as on the symbolic level. This is not to say, I claim again, that these sources' *content* should not be used as evidence. They should, but not so much because it 'represents' an extra-literary reality, but rather because they are *translations* of experiences from one medium into another, undertaken by people considered to be specialist in 'verbalizing' such experiences. Such content often cannot be found in the more traditional sources such as archives and trade journals. In doing so, I am constantly aware that "experience can never be reduced to processes of representation."[24]

If a comparison between fictional and nonfictional media is possible (for instance when social scientific research exists on the same topic in the case of contemporary, nonhistorical research, or when nonfiction accounts can be found in historical sources, as is the case in Chapter 1), the limits of the former can be neutralized. Thus the ubiquitous 'flight' metaphor in both cycling and autopoetic fiction can be explained as a very strong adventurous (and transcendental) element, considering the condition of the roads, which, as we know from more mundane sources, were mostly too bad to enable a smooth ride. Similarly, we can relativize the focus on the individual self in such fiction as a structuring element of the novel and the short story, knowing from the usual sources about the collective (group and family) character of many cycling and automotive touring practices. The recent revival of literary mobility studies tends to ignore or neglect the historical context of fictional utterances suggesting an idyllic culture (even if admittedly "sensationalized" for "literary effect") incompatible with nonfictional historical sources.[25]

The problem with *historical* literary sources about the adventurous aspects of automobilism, however, is that such comparisons with non-fictional sources are often not possible, especially when sociocultural norms inhibit talking or writing in public about such issues. This is for instance the case with the phenomenon of 'inversion' (see Chapter 1), where only a *close reading* of the literary sources allows us to observe a fundamental narcissistic trait of early automobilism: nobody in the contemporary press self-diagnosed as narcissistic; such conclusions are to be found in cultural histories of the fin-de-siècle neurasthenia debate.[26] Geographer-historian Peter Merriman has asked, in a polemic against the recent call for more 'mobile' methods of social science research, for our attention to what his opponents have called 'conventional' or 'traditional' methodologies. At the same time he mobilizes non-representational theories, without disclosing how he intends to analyze historical 'representations' of car culture in a 'non-representational' way.[27]

Non-representational theorists have emphasized that although they "do not prioritize the role of representation in their accounts of the social and the subject," they are not "anti-representation *per se*," so some of them would prefer their epithet to be changed into "more-than-representational" theories. Representation is about "transformation and difference" (a process we have called 'translation') and as such non-representational theory does not "refuse" it, only if it is seen as "the repetition of the same or representation as a mediation."[28] A representation, from this perspective, should be seen as a practice, a "presentation" itself, as a medium "on its own terms": "A face is painted. A sunset is photographed. A room is described. What has already taken place in one medium passes over into what takes place in another medium . . . Non-representational styles of thought treat everything usually regarded as representational (e.g. words, concepts, ideas, perceptions, and images) as events in their own right."[29] For literary theorists this must sound quite familiar (as the 'relative autonomy' of art, a point of view ushered in before the emergence of the sociology of literature), although non-representational theories emphasize the embeddedness of such 'events' in social 'life.' Despite Nigel Thrift's belief in the performing arts that can "jolt people . . . into places they never thought they'd get," *historical* autopoetic fiction remains elusive for non-representational theories, their vagueness (as an openness, for others to build on) seen as a virtue by its proponents.[30] Perhaps one avenue around the contradiction of addressing representations from the viewpoint of practice theory is to consider fiction à la Michel de Certeau as "a form of modern ethnography," a space "populated by everyday virtuosities that science doesn't know what to do with."[31] There is indeed, *pace* Nigel Thrift, a lot "going on that talk cannot grasp," but if we don't have the luxury to 'grasp' such 'absent present' phenomena in contemporary 'reality' through non-representational means, we are lucky to have historical sources that 'talk' about the 'untalkable,' that try in all kinds of registers (novels, short stories, songs, movies, games) to translate the 'untranslatable.'[32] Especially when aiming to 'represent' subaltern historical actors, all forms of scholarly 'tools' should be mobilized, all the more so if we acknowledge that even Deleuze and Foucault "declare that the oppressed subject of history can speak through the intellectual, that is to say, without mediation."[33]

This book presents mobility history as a process of all kinds of movement (of people, things, ideas, bodily gestures) in a period, a 'very long twentieth century,' dominated by the practice of and the discourse about *automobilism* (or in Brigitta Frello's terminology: movement [on and beyond the road] conceived as both empirical and metaphorical),[34] constructed and constantly changed by riders, drivers, passengers, and many other personal and institutional stakeholders, such as governments, experts, non-users (in particular pedestrians and cyclists), and sometimes even technologies, as agency. Apart from the passenger car, the bus, and the truck, two-wheelers (motorized and non-motorized) also play leading roles in this narrative, next to supporting roles for railways, tramways, and very rarely ships and airplanes. Although some mobility scholars emphasize the 'system' character of 'automobility' (the latter a term, borrowed from transport science, that we try to avoid, preferring instead the neologism 'automobilism' [French '*automobilisme*'] as suggesting process and change rather than rigidity),[35] my main goal in this book is to understand the history of the *practice of moving-by-car* (and by bicycle, and by foot) in a *systemic context*, especially the (directly user-relevant) infrastructural context of road network building. One of the main plots in this narrative is the gradual shift, during this period, from an Atlantic to a Pacific stage. The Atlantic, defined initially as an anti-communist civilization in its Northern part in a Cold War setting, experienced a metamorphosis in the masterful hands of later historians who introduced a Southern part of massive trade (of enslaved people and other 'goods') and a no less massive tradition of migration, excellently reviewed by Bernard Bailyn.[36] The Pacific, with an even older history of 'civilizations' (home of "hundreds of distinct tribal societies and languages that flourished in jungles and mountain valleys, and across beaches from the Solomons, Vanuatu, and New Caledonia to Papua New Guinea," often invisible in world histories),[37] has only recently been revived by historians and social scientists alike as the Atlantic's rival in a geopolitical sense, the result of a century-long process of globalization, in which automobilism, and mobility in general, has played and is still playing a dominant role, especially after the energy crises of the 1970s, accelerated even further in the 1990s, "after the end of the bipolar world," and further still around the turn of the last century.[38] This role of (auto)mobility is important, but not "ubiquitous," as several scholars have emphasized who have warned (like Kate McDonald) against the "obsession with mobility and movement" among globalization students to the detriment of place-based histories, and of people with historical agency who threaten to disappear behind abstractions of 'flow' and 'circulation.'[39]

A world mobility history can be written from several angles: institutional, technological, sociocultural. The world mobility history presented here has elements from all three angles, but it concentrates on the bicycle's and the car's *user cultures*, rather than their production, on *conceptual transfers* (for instance of the freeway concept over the globe) rather than their technology, on its *sociocultural differences* rather than its 'system.' This is not to say, of course, that an encompassing world mobility history should not include a history of car technology and repair (introducing labor history, largely absent from the current narrative, also needed to better understand the proliferation of 'informal transport'), or the history of oil production (and related geopolitical aspects of world mobilities). Nor should an 'entangled' or 'connected history' of an emerging global automotive economic 'network' be absent in a comprehensive global mobility history, if only to explain the initial weak position of Chinese negotiators in their effort to build a national industry vis-à-vis the car multinationals (see Chapter 4). Nonetheless, I believe the present *focus on sociocultural factors in a context of technical, economic, and institutional influences* is justified because of their being, in exactly this composition, a major 'driver' of change: perhaps the best example of this claim is the 'victory' of the internal combustion engine car over the electric car at the start of the twentieth century, as discussed in Chapter 1, a victory that still haunts the current efforts to radically change the automotive paradigm as addressed in Chapter 5. However, at this moment in the historiography of world mobility the most important needs are contributions from scholars who read and speak the languages of the countries that here have been accessed only through (highly selective) translations, as well as the dozens of countries that have not been covered at all. Only then would world mobility history have become not only in its products, but also through the community of its producers, a mature discipline.[40]

In Chapter 1, then, we start our analysis of nearly two centuries of mechanized road mobility with the urban rickshaw in Asia, an original 'bricolage' of indigenous and Western technologies and practices, followed by a focus on the Western emergence of the car and its culture from earlier bicycle, carriage, and engine technologies and its 'struggle' with two competing options, steam and electric propulsion. Around this car, an aristocratic and bourgeois mobility culture was built, soon extended with and dominated by a middle-class leisurely culture, supported by national governments and interconnected through transnational associations.

Chapter 2 continues the rickshaw narrative, following its quantitative Asian blossoming and its subsequent demise, caused, depending on the country, by the bus, the jitney, or the car. This chapter also zooms out for the first time to the *entire* globe, analyzing the struggle between road and rail in Africa and Latin America, while in the West national 'car systems' were established by the end of the Interbellum, including a countrywide car-friendly road system.

Chapter 3 sees these systems develop into exuberant Western cultures on freeway networks, a principle that then becomes the subject of an active rollout over the globe, a globalization stimulated by the World Bank and International Monetary Fund, and supported by Western consultancies. Whereas Latin American countries see their middle classes motorize, and Japan promotes its own specific form of automobilism in the Pacific, the Soviet Union sets on the railways for its modernization, and invites China to do the same. In Africa a new phenomenon emerges: the informal semipublic transport of the jitney, a revival of forgotten alternatives in the West.

Globalization governs Chapter 4, in the midst of growing concerns for the environment and ever increasing energy use and pollution, trends that provoke the emergence of new generations of mobility students producing the concept of the 'mobility turn.' We witness the explosion of Asian automobilisms, accompanied by growing concerns about traffic unsafety, monitored and redefined by the World Health Organization as an epidemiological phenomenon, as an issue of public health. In the new century, a re-electrification of the global car fleet seems underway, exacerbating the problems caused by mobility injustice and mobility inequalities.

Chapter 5 is new. It addresses the hope for an 'end of automobilism' and investigates the alternatives to the car society, contrasting global car cultures of carnivalesque mobilities versus status, of autonomous vehicles versus non-motorized alternatives in the cities. We look at the history of car futures and wonder whether the electric car will, after more than a century, lose its stigma of being a perpetual car of tomorrow. The current crisis has also triggered an explosion of suggestions of a culture 'after the car,' a condition to be reached by an activism 'against the car.'

At last, the Conclusions, after a wrap-up of the previous chapters, reflect on what a world-historical approach has brought us beyond what a national-historical or a purely social-scientific approach have enabled so far.

Notes

1. Gijs Mom, *The Electric Vehicle: Technology and Expectations in the Automobile Age* (Baltimore: Johns Hopkins University Press, 2004).
2. Gijs Mom, *Atlantic Automobilism: Emergence and Persistence of the Car, 1895–1940* (New York: Berghahn, 2015); *Globalizing Automobilism: Exuberance and the Emergence of Layered Mobility, 1900–1980* (New York: Berghahn, 2020); *Pacific Automobilism: Adventure, Status and the Carnival of Mobility, 1975–2015* (New York: Berghahn, 2022).
3. Direct quotations are indicated by double quotation marks; single quotation marks either mean 'so-called' or are quotations from my trilogy that, because of space, are not separately justified. The reader is invited to consult the chapters in the trilogy referenced in general at the beginning of the passage concerned.
4. For instance Wladimir Sgibnev, "Post-Socialist Public Transport Development Trajectories: An Overview," oral presentation (2024 T2M Annual Conference, "Mobilities and Infrastructures: Transitions and Transformations," 23–26 September 2024, Leipzig [Leibniz-Institut für Länderkunde and Lancaster University Leipzig]).
5. Sanjay Subrahmanyam, "On the Origins of Global History: Inaugural Lecture delivered on Thursday 28 November 2013; transl Liz Libbrecht," 26, 28, https://books.openedition.org/cdf/4171 (accessed 12 February 2025).
6. Roland Wenzlhuemer, *Globalgeschichte schreiben: Eine Einführung in 6 Episoden* (Konstanz: UVK Verlagsgesellschaft, 2017), 9–23.
7. Dominic Sachsenmaier, *Global Perspectives on Global History: Theories and Approaches in a Connected World* (Cambridge: Cambridge University Press, 2011), 48, 51, 78, 133–34; Sebastian Conrad and Andreas Eckert, "Globalgeschichte, Globalisierung, multiple Modernen: Zur Geschichtsschreibung der modernen Welt," in Sebastian Conrad, Andreas Eckert, and Ulrike Freitag (eds), *Globalgeschichte: Theorien, Ansätze, Themen* (Frankfurt: Campus, 2007), 7–49, here 18, 23, 26; on deviation: Emily Apter, *Against World Literature: On the Politics of Untranslatability* (London: Verso, 2013), 157.
8. Helmuth Berking, "Raumtheoretische Paradoxien im Globalisierungsdiskurs," in Helmuth Berking (ed.), *Die Macht des Lokalen in einer Welt ohne Grenzen* (Frankfurt: Campus, 2006), 7–22, here 9.
9. Patrick Manning, *Navigating World History: Historians Create a Global Past* (New York: Palgrave Macmillan, 2003), 222–23; however, see also his "Global History and Maritime History," *International Journal of Maritime History* 25, no. 1 (June 2013), 1–22; Daniel R. Headrick, *The Tools of Empire: Technology and European Imperialism in the Nineteenth Century* (New York: Oxford University Press, 1981).

10, Arnold Pacey and Francesca Bray, *Technology in World Civilization: A Thousand-Year History; Revised and Expanded Edition* (Cambridge, MA: MIT Press, 2021 [1990]), Chapter 10 (195–216, covering the period 1860–1960).
11, Nigel Thrift, Adam Tickell, Steve Woolgar, and William H. Rupp (eds.), *Globalization in Practice* (Oxford: Oxford University Press, 2014). The two chapters are "Road Safety and Traffic Management" (79–83, written by Daniel Neyland and Steve Woolgar) and "Mobility" (31–34, written by Peter Merriman).
12, John Urry, "Globale Komplexitäten," in Berking (ed.), *Die Macht des Lokalen*, 87–102, here 92; McNeill quoted in Sachsenmaier, *Global Perspectives*, 74.
13, Arnaud Passalacqua, "Mobility Techniques: Between Globalization and Local Solutions," in Guillaume Carnino, Liliane Hilaire-Pérez, and Jérôme Lamy (eds.), *Global History of Techniques (Nineteenth to Twenty-First Centuries)* (Turnhout: Brepols, 2024), 351–62.
14, Kate McDonald, "Imperial Mobility: Circulation as History in East Asia under Empire," *Transfers* 4, no. 3 (Winter 2014), 68–87, here 68, 83.
15, Pernau quoted in Conrad and Eckert, "Globalgeschichte," 39; Anna Wierzbicka, *Imprisoned in English: The Hazards of English as a Default Language* (Oxford: Oxford University Press, 2014).
16, Apter, *Against World Literature*.
17, Mom, *Globalizing Automobilism*, 38–42; for one of the first studies to include the 'mobility turn' see: Anita Perkins, *Travel Texts and Moving Cultures: German Literature and the Mobilities Turn* (Oxford: Peter Lang, 2016), 217.
18, Mom, *Pacific Automobilism*, 14–18.
19, Mom, *Globalizing Automobilism*, 35–36. For a scholarly 'call' to integrate mobility and media studies that did not acquire a following, see: Dorit Müller and Heike Weber, "'Traffic'—On the Historical Alignment of Media and Mobility," *Transfers* 3, no. 1 (Spring 2013), 65–74.
20, Mom, *Atlantic Automobilism*, 3–7.
21, See the founding of the journal *Mobility Humanities* (2021) by the Academy of Mobility Humanities of Konkuk University in Seoul, South Korea. See also the founding of a Mobility & Humanities Centre of Advanced Studies at the University of Padova (https://www.mobilityandhumanities.it, accessed 24 February 2024).
22, Nigel Thrift, *Spatial Formations* (London: Sage, 1996); see also Nigel Thrift, Paul Harrison, and Ben Anderson, "'The 27th Letter': An Interview with Nigel Thrift," in Ben Anderson and Paul Harrison (eds.), *Taking-Place: Non-Representational Theories and Geography* (Farnham: Ashgate, 2010), 183–98.
23, Mom, *Atlantic Automobilism*, 34–40.
24, Mom, *Globalizing Automobilism*, 28–30; last quote: Ben Anderson and Paul Harrison, "The Promise of Non-Representational Theories," in Ben Anderson and Paul Harrison (eds.), *Taking-Place*, 1–34, here 204. On the problematic relationship of non-representational theory with history ("memory") see ibid., 204–5.
25, See for instance Una Brogan, "Cycling and Narrative Structure: H.G. Wells's *The Wheels of Chance* and Maurice Leblanc's *Voici des ailes*," in Marian Aguiar, Charlotte Mathieson, and Lynne Pearce (eds.), *Mobilities, Literature, Culture* (Cham: Palgrave Macmillan, 2019), 237–57; on "sensationalized": Peter Merriman, "Rethinking Mobile Methods," *Mobilities* 9, no. 2 (2013): 167–87, here 181.
26, Joachim Radkau, "'Die Nervosität des Zeitalters,' Die Erfindung von Technikbedürfnissen um die Jahrhundertwende," *Kultur & Technik* 3 (1994): 51–57; Gijs

Mom, "Civilized Adventure as a Remedy for Nervous Times: Early Automobilism and Fin de Siecle Culture," *History of Technology* 23 (2001): 157–90.

27. Merriman, "Rethinking Mobile Methods," 168, 173.
28. Anderson and Harrison, "The Promise of Non-Representational Theories," 2n1 ("prioritize"), 19 ("*per se*"; italics in original), 25 ("refuse" and "mediation"). On 'translation' see Mom, *Globalizing Automobilism*, 28–38.
29. Marcus A. Doel, "Representation and Difference," in Anderson and Harrison (eds.), *Taking-Place*, 117–30, here 119–20.
30. Thrift, Harrison, and Anderson, "The 27th Letter," 195–96.
31. Ben Highmore, *Cityscapes: Cultural Readings in the Material and Symbolic City* (New York: Palgrave Macmillan, 2005), 114 (the first quote is Highmore's).
32. Nigel Thrift, "Intensitäten des Fühlens: Für eine räumliche Politik des Affekts," in Berking (ed.), *Die Macht des Lokalen*, 216–51, here 222.
33. Tariq Jazeel and Stephen Legg, "Subaltern Studies, Space, and the Geographical Imagination," in Tariq Jazeel and Stephen Legg (eds.), *Subaltern Geographies* (Athens: University of Georgia Press, 2019), 1–35, here 15.
34. Birgitta Frello, "Towards a Discursive Analytics of Movement: On the Making and Unmaking of Movement as an Object of Knowledge," *Mobilities* 3, no. 1 (2008): 25–50, here 29.
35. See, for instance, John Urry, "The 'System' of Automobility," *Theory, Culture and Society* 21, no. 4/5 (2004): 25–39.
36. Bernard Bailyn, *Atlantic History: Concept and Contours* (Cambridge, MA: Harvard University Press, 2005).
37. Matt K. Matsuda, *Pacific Worlds: A History of Seas, Peoples, and Cultures* (Cambridge: Cambridge University Press, 2012), 3.
38. Giovanni Levi, "Frail Frontiers?" *Past and Present* (2019), Supplement 14: 37–49, here 38 ("bipolar"); Richard Drayton and David Motadel, "Discussion: The Futures of Global History," *Journal of Global History* 13 (2008): 1–21, here 6.
39. Sebastian Conrad, *What is Global History?* (Princeton, NJ: Princeton University Press, 2016), 226; Stefanie Gänger, "Circulation: Reflections on Circularity, Entity, and Liquidity in the Language of Global History," *Journal of Global History* 12 (2017): 303–18, here 312 (agency).
40. Gijs Mom, "New Mobility Studies: Niche to Mainstream; History and Social Sciences in the Mobility Turn" (*Transfers*, forthcoming).

CHAPTER 1

Imperialist and Colonialist Mobilities, 1850–1918

A world history of road mobility in the nineteenth to twenty-first centuries should start with imperialism and colonialism and mobility's role therein. It should also start by decentering the West, so why not take Asia, for instance British 'railway imperialism' in India and a more comprehensive 'mobility imperialism' in Korea and China during Japan's Meiji Restoration (1868–1912), a 'hybrid modernization' period characterized by a layering of the old and the new, mercantilism and monopoly capitalism, driven to a large extent by mobility intensification, shattering "the illusion that industrialization was a uniquely Western process"?[1]

Imperialism's primary mobility modes, its 'tools of empire,' were shipping and railways, both instrumental in the emergence of transnational migration as well as tourism. Whereas in Africa, especially after the Berlin conference of 1885, railways were constructed as 'imperialist' through their extractive and military function, leaving a non-unified patchwork of lines from coasts to mines and areas of rebellion, in India the colonizer created "the most monumental project of the colonial era," representing "the largest international capital flow of the nineteenth century," resulting in one of the longest networks in the world, after the US, Canada, and Russia. Here, too, "competitive, imperial strategic thinking" (such as serving the Lancashire 'cotton barons' and military arguments, especially after the Indian Revolt of 1857) were the development motives, but later in the century anti-colonial people could use the network to their own benefit, opposing a 'European' system of colonialism that by 1914 occupied 84 percent of the world's land area.[2]

Steam (shipping and railways) crucially helped create a "world-economy not so much by trade flows as by more or less unilateral transfers of labor, capital and entrepreneurship flowing from the rest of the world to its political jurisdictions."[3] Lately, this 'top-down' style of history writing has been questioned by a new generation of historians who wish to go "beyond what may be described as [the] 'Headrick-Adas model,'" and instead aim to give a (culturally informed) "view from below," claiming (and quoting railway history doyen Colin Duvall) that "transport history continues to ignore users."[4] Although this is certainly not the case for road mobility history (which is, in this respect, tributary to a library on 'user agency' in the history of technology), railway histories also seem to have been 'catching up' with user culture, including those on India.[5]

In the century before 1920 "more than one-half [of] the European populations moved internally," whereas the Atlantic world saw 50–55 million people and the Asian world (perhaps more than) 30 million people migrate. In less than one century, more than 150 million people left their homes to go elsewhere.[6] Less than 2 percent of the world population was constantly on the move during the surge of migration by the end of the nineteenth century, pulled and pushed by developments like "new information technology, conflicts over natural resources, and the consolidation of international regulation," but also by developments that certainly did not occur in the West alone, as an Orientalist reading of global migration would have it.[7]

British firms built the first railway lines in Japan in the 1870s. When the victory in its war with China, fought in 1894 in Korea, gave Japan the control of Taiwan and the railroad building rights in southern Manchuria, and its victory in the war with Russia in 1904 provided the control of the Russian railroads in Manchuria, followed by the outright annexation of Korea in 1910, the colonial railway lines were constructed by Japanese engineers, not only for tourism but also for the extraction of resources. Korea's fledgling railway network is an example of the need to avoid applying Western railway-historical 'common sense,' as not only did it differ from the usual Western imperialist railway network construction in that the first line was a north–south trunk line (forming the spine of a later hinterland network), the country also had a centralized water and land transport network around Seoul, which then became centered around the coastal ports and the newly built and militarily inspired Japanese railways, meant from the beginning to invade China and Manchuria.[8]

Japan's *urban modernization*, however, not only took place on rails, in the form of local tramways, but also on the streets, where during the Tokugawa period (1615–1868) wheeled transport was forbidden and only water and pedestrian transport (including the sedan chair or palanquin) were allowed. Bicycle technology was one of the first to claim a place in Japan's modern mobility culture, but the real explosion of a rather utilitarian user culture driven by the working classes' commutes occurred after the introduction of the 'safety' bicycle (see below) in 1889: by the end of the Meiji period in 1912 Japan counted 400,000 bicycles, only a couple of hundred among them ridden by women. In 1908 the number of bicycles had surpassed that other iconic vehicle, the rickshaw.[9]

Invented shortly after the beginning of the Meiji period, the rickshaw or *jinrikisha* (man-powered vehicle, based on bicycle technology with its parallel spoked wheels and lightweight structure) became an instant success: by 1875 one hundred thousand of these hand-pulled two-wheelers were plying Tokyo's streets. The production was truly transnational: materials came from India, Britain, Germany, China, Spain, and Japan. From 1904 its wheels were equipped with solid rubber tires, four years later replaced by pneumatics, lowering the traction resistance considerably and making the use of a 'pusher' superfluous.

It was the diffusion throughout Asia (and beyond, to the African city of Durban, for instance, where by the end of the nineteenth century 2,000 had been imported from Japan) and its local appropriations that made the rickshaw into a crucial tool in the livelihood of rural–urban national and transnational migrants, their subalternity still suffering from "relative [scholarly] neglect."[10] It probably started when in Shanghai a French businessman imported 300 in 1874, and five years later the local fleet had expanded to 2,500. The first ten rickshaw fleets were foreign-owned, but from the 1880s ever more rented rickshaws fell into the hands of the Green Gang. A traffic survey in the International Settlement as late as 1918 still counted nearly 15,000 rickshaws and only 722 bicycles and 1,863 cars (and no sedan chairs). Half of the fleet of about 20,000 was in private hands.

Whereas the size of the Japanese domestic fleet started to decline from 1910, export peaked a decade later, shortly before the factory was destroyed by the Great Kantō Earthquake of 1923. It seems that a substitution process was responsible for this: from shortly after the turn of the century the horse-drawn tram, followed by the electric version, and the bicycle, started this process (nearly halving the number of renter-pullers), while the appearance of the motorized taxi and the bus in 1912/13 reduced the number of owner-pullers further. Yet by 1910 hardly a hundred cars were plying the streets of Japan's capital, so a competition between rickshaws and cars did not play a role, at least not in Japan. Nor was the car able to replace the ever growing fleet of bicycles, making it historically questionable to depict the bicycle as a universal 'predecessor' of the car.

Indeed, early automobilism in Japan resembled the largely utilitarian rickshaw and bicycle culture. In Kyoto, for instance, the first cars were run "as a kind of small bus," and the earliest indigenous car, called the Dat (in Japanese pronunciation Datto, meaning "scooting rabbit"), was small and slow, often (because of the lack of a mass market) deployed as "hire cars." Such tendencies, to see the car primarily as a collective vehicle for semipublic use, were not unknown in the West (see, for instance, the booming [often Black] jitney culture in the 1910s and 1920s in the US and Canada), but there this culture was by then already overshadowed by individualized car hedonism: its zenith lay in 1915 when the US counted 62,000 privately owned and privately used cars, and Canada around 3,600.[11] Nonetheless, it is worthwhile to look at the 'West' from the 'East' as it helps uncover unexpected mobilities, such as Afro-Americans who as early as 1905 had to boycott one of the first American bus companies, and Black investors in Nashville who "pooled their assets to buy a fleet of automobiles 'to furnish a passenger service where there will be no lines drawn between the whites and the Blacks.'" It seems that even Greyhound Bus started as a white jitney service in 1914 in Minnesota.[12]

By far the most detailed and thorough analysis of the culture of the collective rickshaw communities has been given by Australian historian James Warren, whose 'historical ethnography' of the mobility culture of Singapore was later often used, unjustifiably so, as a 'model' to describe fleet cultures elsewhere in Asia.[13] Largely based on coroner's reports on the causes of pullers' deaths, Warren's study showed that most of them were poor migrants from the southern provinces of China, who rented their vehicles on a daily basis from fleet owners with the same Chinese background. Within one year of their introduction in 1880, one thousand were counted in the city state, but it is not clear whether the vehicles were imported from Japan or locally manufactured. In 1902, 20,000 pullers (all men) were officially registered, who during their strike a year later showed the city's dependence on their service: three-quarters of the Europeans relied on the rickshaw for their commute, a function dominating in the vehicle's use profile, above social visits and going to the market. In Singapore, the rickshaw was stimulated (instead of threatened, as in Tokyo) by the tram (electrified in 1905), and only started to decline in the early 1920s (when the fleet size peaked at 30,000). In 1917 a traffic census on a bridge counted an equal amount of rickshaws and cars, but again: historical proof of a direct substitution does not exist. Nonetheless, such jewels of historical evidence are a more reliable assessment of the car's (un)importance in the daily lives of urbanites than the popular search for the 'first car': Java 1894; Singapore 1896, where the first motor traffic regulations date from 1907; Bangkok 1897. Yet such dates are not unimportant: they suggest a fundamental synchronicity in worldwide mobility cultures, rather than the reflex to see the Global South as haunted by a 'cultural lag.' This also applies to car culture's 'dark side': the first dozen traffic deaths were counted in Indian Calcutta in 1908.[14] The same goes for the bicycle: in West Africa in 1911, anthropologist Hans Peter Hahn showed, Africans and Europeans alike practiced cycling as an "acknowledged sport," whereas in colonial India telegram delivery boys and postmen used the bicycle before 1914 and the Indian middle class had put up cycle clubs and races.[15]

China's mobility modernization followed a different path. There, railway concessions were acquired by France, Germany, and Russia in the second half of the 1890s, and the nationalization by the Qing court may have been one of the triggers of the 1911 revolution, which made medical doctor Sun Zhongshan (in the West known as Sun Yat-sen) into the National Railway Supervisor of the new republic, after he had visited several large Japanese cities. By 1918 about 7,000 km of railway lines had been built.

The rickshaw appeared in Beijing in 1896, and evolved rapidly into a modern vehicle with its ball bearings and light frame. The middle class (shopkeepers, teachers, minor officials) rather than 'the rich' were the dominant patrons, and most pullers were semiliterate city people rather than rural migrants (including some women), "poor but not impoverished," according to David Strand's excellent study.[16] Together with the one-wheeled barrow (also used for passenger transport for up to eight persons who through their seating arrangement kept the vehicle in balance), the rickshaw not only fitted seamlessly in the inner cities with their narrow, meandering streets, but its collective and at the same time individualized character also seemed to resonate better with local demand of a large variety of clients as well as being a welcome albeit harsh livelihood of the subaltern.

As to the alleged 'cultural lag': for Asians the West was at the same time close and far away. As one of many visitors to the West supported by newspapers and the Japanese government, Natsume Sōseki (1867–1916), considered to be "one of Japan's greatest writers," visited London from 1900 to 1902, where he not only lived "like a poor dog that had strayed among a pack of wolves," he also learned to cycle (on a lady's bicycle, because of his small body size) and became aware of the color of his skin.[17] At the same time, the West was far away: in 1915 Sōseki, back in Japan, confessed he was fully unaware of the course of the war in Europe.

In his fiction, pedestrianism, rickshaws, and other animal-powered vehicles are everywhere in Tokyo's layered mobility, but the 'modernist' (or better: modernized, in London) observer had apparently learned how to look at (and look for) them: in front of a house, one protagonist spots "a couple of carriages but no automobiles." When in 1909 he visited Manchuria (by ship and train) upon invitation from the colonial railroad company, he used similar condescending qualifications about local rickshaw pullers as those that had him feel his yellow skin for the first time back in London.[18]

Sōseki was in London at the end of the so-called bicycle craze of the second half of the 1890s. Because we know so much about this culture, it is worthwhile to self-reflect on the way we analyze its emergence and tell its genealogy, all the more so because it can help us to improve our analysis of mobilities during the remainder of the century.[19] First of all we assume that we only can understand a phenomenon through its history, the *changes* it went through to become what we wish to analyze. Further, we should know that '*the* bicycle' was not 'invented' at a certain point in time by a certain 'inventor'; instead, we observe a *continuity* of the gestation of many different really existing bicycles, starting with the *vélocipède* somewhere in the 1810s, used as an ostentatious plaything by the French aristocracy. Half a century later, this idea of a human-propelled two-wheeler was revived when two French mechanics added a crank and pedals, and it was this concept that became the object of experimentation in all kinds of directions, first and foremost in the UK as the most industrialized country at the time. The abstract 'typical' term 'bicycle' is not adequate to describe this process: while several subtypes appeared to be dead ends, some received an ever larger front wheel and an ever smaller rear wheel allowing acrobatic and speedy riding (they were called 'ordinaries'). Others were given three or four wheels, or two saddles in 'tandem,' for those who wanted to roam the countryside rather than race, or carry an extra passenger on their leisurely touring. Whereas the former were used by mostly young male daredevils, the latter were used by the quieter folk, stereotypically understood as women and more timid men.

The bicycle craze started in full when the British J. K. Starley (1854–1901), nowadays celebrated in popular publications as 'the inventor' of 'the bicycle,' introduced in 1885 the 'safety', a bicycle type with two wheels of equal height and a tubular diamond-shaped framework. It was further developed by the industry (especially Rover) into a lightweight contraption (tubular frame, friction-reducing ball bearings, wire-spoked wheels) and equipped with pneumatics, reinvented in 1888 by the Brit John Boyd Dunlop (1840–1921).

Of course, this story can be extended with developments in other countries, to create a transnational history. We will do so when we deal with its cultural aspects, because we prefer to deal with technology from the viewpoint of the user, but for the moment the previous, somewhat linear prehistory and the multifaceted final two decades after 1875 will do for our argument. What is crucial in this narrative is that the ordinary was not an aberration from a linear development path of two-wheelers (from the *vélocipède* to the safety); instead, it was crucial because it introduced 'speed' into the bicycle's genealogy, by literally integrating this into the safety's structure, as 'safe speed,' never to leave the bicycle ever after. Thus, the relatively light safety on pneumatics 'won' the battle of the alternatives because of its multifunctionality: as a 'fast' vehicle, it also afforded the practice of the former three- and four-wheeler and several other functions, such as touring outside the city, as well as figure-riding and other semi-acrobatic skills in the arena.

⟦ The story just told can be called quasi-evolutionary as it describes the 'struggle' of the different bicycles to develop their technical *properties* such that they 'afford' (in the terminology of the psychologist James Gibson, who in his turn based his approach on the 'action theory' of Theodore Schatzki) certain *functions* ('quasi' because artifacts do not copulate; they 'evolve' through the actions of people).[20] This *practice-theoretical approach*, called 'the dual nature of technology,' is tributary to the history and philosophy of technology and innovation studies. According to this approach, properties do not prescribe functions, as some social scientists claim, using the concept of 'script':[21] the relation between properties and functions is fuzzy, the former realized by the 'producer' (in her expectation of a certain practice by the 'user'), the latter by the 'user' (by approving or disapproving of the work of the 'producer' in her 'buying' behavior and her 'use profile,' the activities she intends to realize in the use of the artifact).

Another possible approach (the social construction of technology, or SCOT) stems from the social sciences, and became very influential in the 1990s and beyond, also in the history of mobility, especially because one of its founders (Wiebe Bijker) used the history of the safety as a case study.[22] SCOT set in motion a rethinking of the design and production process of technological artifacts by including several stakeholders such as 'relevant social groups' of users of these artifacts, in this case 'sporting' young men and women, the first opting for the 'ordinary', the other (later joined by the first) for the 'safety.' One of the points of critique was that the theory was inspired by relativist philosophies and thus lacked a moral and political point of view.[23] But despite its rather reductionist approach (by schematizing a complex historical process into a dichotomy of opposing groups), what SCOT accomplished was a theoretical debate in a rather theory-poor historical profession. SCOT practitioners exposed an older (for many old-fashioned) branch of historians, accusing them of being 'antiquarians,' 'naïve empiricists,' and 'Whiggish historians.' This latter group in turn accused the 'theorists' of 'distortion of the historical account' by making 'conspicuous factual errors' in order to support their theory (caused, they charged, by the use of questionable sources such as amateur histories). ⟧

The safety, then, should not (or not only) be characterized as a 'socially constructed speed machine,' but as a (socially constructed) 'adventure machine,' a difference that becomes clear when we look at *transnational* bicycle *culture*. According to one of its best analyses, bicycle culture emerged in the pivoting period of the fin de siècle: it was 'euphoric' in that it united the mental tendencies of the era into a strange mixture of a rational perception of technology and romantic feelings of the reform movement.[24] Through its practices of speed, touring, and figure riding, the culture should be contextualized as a sport, in the tradition of horseback riding, but its mechanical aspect was new: in order to practice the sport one had to acquire some basic knowledge of the device in order to maintain it, and one had to be prepared to stop at unexpected moments for repairs, especially of the fragile pneumatic. Thus, the bicycle adventure was tripartite: it encompassed the thrill (and skill) of riding at a speed that was considered 'fast' (15–20 km/h), the uncertainty in navigating the countryside, as well as the challenge (and skill) of mastering the technology during maintenance and repair. 'Flight' seemed to characterized the 'movement' in a double sense: flight from the city into 'nature,' while at the same time indulging in the lightweight 'poetry of motion' celebrated in the accounts of bicycle trips and clearly contrary to the practice necessary to negotiate rough road pavements.

From the early 1890s the 'movement' (because as such it was experienced)[25] split into two institutional domains, that of the races and that of the touring clubs; in between a more 'domestic' culture of trips and touring developed. To get a grip on the collective social and cultural practices of the institutional touring community, we can rely on the myriad reports of individual and group experiences sent to the clubs' periodicals.[26] But in order to understand the cyclists' motifs, motives, and motivations, one type of source stands out in this respect, written by a new type of literary writer, self-consciously being a pioneer in both domains: in the practice of riding and the practice of writing. One of these pioneers was a woman, Mrs. (Mary) Edward Kennard, who had written some conventional novels on horse culture, but then decided to shift to the New Woman and her dealing with the nuts and bolts of cycling, in two books, including *Guidebook for Lady Cyclists* (1896). Women's entry into bicycle culture inspired novelists to play with gender roles, such as the American Willa Cather in her short story "Tommy, the Unsentimental" (1896), whose androgynous protagonist raced against an electric tram and competed with and saved 'weak men.'

There were differences between national cultures, however. The pivotal writer for the British cycling culture is widely considered to be

H. G. Wells, not only because he was a cycling pioneer himself, but also because of his lower-middle-class background, which he shared with an increasing contingent of the British cycling movement, according to Yoonjoung Choi. Wells celebrated in his *The Wheels of Chance* (1896, and not *The Wheels of Change*, as I erroneously wrote in an earlier publication) the democratic class- and gender-breaking character of the new sport (and, often forgotten, age as well; many cycling pioneers were school children). His was a moral tale: he 'feminized' his male antagonist (with his 'fine-made' body and his 'pallid complexion'), and let him timidly flirt with an 'unwomanly' woman, a critique of the disciplined manliness of elite sports such as cricket.[27]

In France, at about the same time, 'dandy' Maurice Leblanc published a serial novel in the magazine *Gil Blas* (where he was known because of his column on cycling), published in 1898 as *Voici des ailes!* (*Look! Wings!*), a book-long eulogy on 'flight' while cycling. An effective combination of high- and lowbrow cultures, the novel became an immediate success with eight editions during its first year, celebrating not only the cyborg character of cycling (the bicycle as a prosthesis of the 'meat body' of the cyclist), but also showing one of the central elements of the mobility adventure: the affinity between the mastering of the machine and the conquest of the woman during a tour through Normandy of several weeks by two hetero couples.

Wells's protagonist Mr. Hoopdriver, riding shakily on his secondhand, near-antique bicycle, has inspired some historians to call the cycling movement "essentially a recreational working-class activity" with a "socialistic quality."[28] It is difficult to accept this assessment without further, more detailed evidence: like Una Brogan's analysis of Wells's work, Choi's does not embed his analysis in the state of the art of the history of cycle mobility, their preference for a theoretical, somewhat dehistoricizing approach with an Anglophone bias thwarting a historical weighing of their argumentation.[29] They are thus compelled to accept their literary sources, including their inherent orientation on the individuality of the 'Self,' at face value, and miss the movement's fundamental collectivist club culture, where many 'outings,' certainly around the turn of the century, and perhaps more so in other European countries, were group outings, their members often clad in special uniforms and led by a club official, a practice derived from military culture. Granted, as vanguards in an early industrializing society, social historians showed how big cities in Great Britain saw a "new breed of petty bourgeois shop and office workers" emerge, soon joined by "schoolteachers and other minor professionals," among them quite a few women, often "single living and commuting in the big city." They were overrepresented in voluntary associations (such as bicycle clubs), "impugned as effeminate as well as unmanly" enthusiast consumers (and producers) of popular culture, and "in the front line of engagement with modernity."[30] Such historians emphasized the "obsessive pursuit of status and respectability" as well as the aggressiveness and manliness of this "new comfortable" middle class (such as rentiers and stockbrokers, but also engineers, authors, male teachers, and office clerks, the most numerous latter's "quintessential desperation" and their "desperate need for adventure" characteristic of the "predominantly male" British sports culture, including H. G. Wells's bicycling Mr Hoopdriver).[31] Although there exists no solid quantitative evidence for this, given the relatively modest costs (compared to other sports such as fishing or hunting)[32] the number of bicycle owners passed the total size of club membership probably in the course of the 1900s.[33] By the end of the remarkably short and astonishingly effective craze around 1900 a new bicycle still cost three times the monthly income of a schoolteacher. Even if the bicycle in its second-hand version became affordable for workers, "the long working hours, lack of paid holidays, and disinclination for physical exercise which normally accompanied a life spent at manual labour" would have made the expansion of the sport beyond the middle class, certainly during the early years of the new century, rather unlikely.[34] Even then outings also became family events, or practices of groups of friends. Whether this expanded user culture also applies to other Western countries with a bicycle movement than the UK is questionable, however, certainly at this early moment, when in every country cycle clubs and individual riders had to cope with fierce (and sometimes violent) resistance in the countryside, as the 'elite' cycling urbanites were disrupting the street culture of children playing and villagers socializing on the street. Uncritically celebrating early bicycle culture is to neglect its basic 'mobility inequality.' Also, the Anglophone bias of cycling historiography tends to block from our view a more elite, more adventurous, more organized and collective culture and (therefore) one more confronted with resistance on the European continent, although David Rubinstein as early as 1977 already drew a picture of a *British* cycling culture haunted by bad roads, hostility from countryside people, and the protective function of collective club tours.[35] There, J. W. Allen as late as 1909 wrote in his *Wheel Magic*: "Every stone on the road, every nail and bit of broken glass, every greasy patch is an ambush. Waggons are waiting for [the cyclist] round corners on the wrong side of the road; cows are preparing to emerge suddenly from invisible gates. Can he [*sic*] trust his chain, can he trust his brakes, above all, can he trust himself?" Remarkably, these uncertainties, including those of possible accidents, "add a charm and a zest to every mile . . . One of the finest qualities of cycling is just that it involves an element of difficulty and even danger."[36]

But it cannot be denied that after the 'craze' of the 1890s the 'movement' grew precipitously, and gave the cycle clubs considerable social and political power in the regulation of traffic and the building of cycle paths, or the cry for 'good roads' in the US. This was less the case for the Arbeiter-Radfahrerbund Solidarität (Worker Cyclists' League, founded 1896) in Germany, a country with 63 percent of the workforce consisting of unskilled workers: they were boycotted by the government and fellow cycling associations, but its membership grew to more than 143,000 in 1912 and even more than 300,000 in 1930. Connected to the Social Democratic Party, their members founded their own cooperative bicycle factory and a savings bank. Opposed to 'capitalist' competition and races, they countered gender stereotyping through organizing indoor "acrobatic cycling" and parades, especially during the 1 May celebration, where their bicycles were decorated with red ribbons and red flowers.[37]

National touring clubs cooperated transnationally in the Ligue Internationale des Associations Touriste (LIAT, 1899). By the end of the first decade of the new century, cycling seems to have lost much of its adventurous character and the utilitarian function of the commute had replaced the original thrills and euphoria, driven no doubt by thousands of cyclists from the middle and lower-middle classes. We have to be careful here, because the adventurous character of cycling in the early years of the Interbellum has hardly been investigated. By then, an evolutionary branching off of the 'safety' had seen an even lighter bicycle emerge, specifically developed for dedicated racing; a dedicated 'feminine' cycle also appeared, with an underslung frame but heavier, to allow women who rejected 'bloomers' to keep wearing their long skirts while riding. Now, adventure became 'delegated' to the racers, including women "circus riders" whose exercises indoors and outdoors drew large crowds.[38] Leisurely touring remained a popular pastime without the development of a specialized 'touring bicycle': the safety would remain a 'fast' vehicle for the remainder of the century, and beyond. The French road safety expert Jean Orselli has calculated that the bicycle (together with horse-drawn transport) dominated in France until the beginning of the 1920s, both in terms of vehicle-kilometers and in passenger-kilometers. Road censuses suggest a similar condition in the UK. In the Netherlands the predominance of the bicycle in passenger-kilometers even lasted until the end of the Interbellum.[39]

Indeed, the history of early cycling should be seen as thoroughly gendered. Women cyclists in general changed the movement fundamentally: their cycling practices, in the public mind related to the suffragette movement in several countries, often 'in masculine clothes,' disturbed many contemporaries. Studies about 'the crisis of masculinity' in this period abound: the bicycle (and later the car) fitted seamlessly in the efforts to overcome this crisis through the (futile) creation of a 'separate sphere' of homosocial bonding in sports in general, but particularly in the 'mechanical' sports of cycling and automobilism. One aspect of the gendered character of early cycling history is a phenomenon often somewhat neglected in the historiography: a dedicated study on the "annual Niagara-on-the-Lake bicycle gymkhana" in Canada, which started in 1896, showed that it drew large crowds, who could witness "a brigade of parasol-laden firefighters on decorated bicycles and a rolling cascade of flowers," practices undertaken intentionally as a statement of "polite bicycle recreation" against the aggression of cyclists who considered it "a common thing . . . to knock people down and then ride away." Such events were a clear cultural statement: a "Tortoise Race" rewarded the cyclist who finished last.[40] Such types of activities will, much later, be taken up again by other 'marginals' such as Blacks and Latinos in the US, and lowriders and car modifiers in Europe and the rest of the world (see Chapter 4).

It is this gendering of cycling that recently attracted new scholarship to the bicycle craze. It allows us to test the reliability of fiction as a historical source. Whereas Una Brogan emphasized the idyllic character of the movement and pointed at the movement's "limits" in bringing about women's "liberation," Eva Chen expanded our knowledge considerably by emphasizing the 'sartorial' aspects of early cycling. By intertwining speed (as a "crucial part of women's experience," testified to, among other things, by women's early participation in cycle racing), with fashion, women pushed cycling more toward a consumer culture, thus embedding cycling more solidly within societal discourse.[41] At first sight Brogan and Chen arrived at these opposing conclusions because they used (or emphasized) different fictional sources, such as Grant Allen's *The Type-Writer Girl* (1897), who rides "recklessly" downhill, "my feet on the rest, and my hands too incautiously removed from the handle-bar," resulting in a collision with another female cyclist. In this novel the bicycle is, like the typewriter, "a machine of speed."[42]

My rereading of Wells's novel confirms that the "adventure" in the subtitle may largely be of a spatial ("wandering") character, but the temporal aspects (speeding, chasing) are certainly not absent, nor its functional sides (a puncture, inspection of the mechanism). Similarly extreme violent fantasies are there ("the blood of the inhabitants squirting this way and that"; "he was angry and adventurous"), whereas the quality of the (white-dusty) roads is generally good, despite the many bricks and other objects spoiling the wish to enjoy the surrounding landscape. The real adventure, however, is to be found in the erotics (Hoopdriver's "knightly quest") of conquering a 'free woman' and the thrill of crossing both a gender and class divide.[43] Perhaps Wells's more radical socialism (than the Fabian Society, which he left because of its lack of activism) incited him to somewhat "propagandist" plots in which not only a kind of "'genteel' proletarianism" shines through, but also a clear class-based car critique, which, by the way, changed into endorsement of automobilism (and some hostility toward cycling) once he bought a car in 1914.[44]

I venture that the controversy about the 'adventurous' character of cycling (and later car) culture is the result of a misunderstanding about its characteristics: both Choi and Brogan situate the 'sporting' and 'scorching' culture in the very early high-wheel phase, thus in fact following Wells himself, who needs this to propose his vision of a 'moderate,' 'rational' culture.[45] The next step in this misunderstanding, then, is to interpret, unhistorically so, coexistence (of adventure and commodity, thrill and status) as substitution, and ignore the fundamental flexibility of the adventure machine in all its temporal, spatial, functional, and violent aspects.

But whereas in Anglophone mobility cultures the bicycle soon would become a 'poor man's vehicle,' in other cultures cycling remained thriving: in Germany, for instance, different from the socialist Clarion clubs in the UK, working-class clubs by 1913 had gathered 150,000 members (and double that amount in 1930). Recent scholarship on cycling's transnational and gendered culture convincingly shows that in every industrializing country there were at least three distinct groups of cyclists: club members (who up to now have dominated historical cycling scholarship, including mine), bourgeois 'domesticators' (mostly women for whom cycling was not so much an emancipatory practice as an effort to encapsulate the culture into a conservative family atmosphere), and between these the heterogeneous group of 'unattached.'[46] Gender and masculinity studies have showed that women's role was important in all three groups, to different effects.

[Gender studies nowadays include men and masculinity studies: the early British (and by extension transnational) bicycle culture cannot be understood without the insight that "Victorian boys learned a feminized version of masculinity from women. Boys and men countered this feminization in numerous ways."[47] Staring at the vehicle (the suggestion that the vehicle invited or enabled 'masculinity') is misleading; instead of a 'script' it is culture that reigns supreme: once the bicycle became 'tamed' such boys, meanwhile somewhat older, got into the car (as Wells did, and many with him), and later they even got in the plane, as we will see later. Adventurousness is not bound to a vehicle.

Since R. W. (Raewyn) Connell in 1995 published her distinction of 'multiple masculinities' (including a hegemonic form, "guarantee[ing] the dominant position of men and the subordination of women"), the study of the role of masculinity in history, including mobility history, has proliferated. One result of this is a thorough insight into the "re-invent[ion of a] muscular form of outdoor, imperial masculinity" by middle-class white men at the turn of the twentieth century.[48] Another exemplary analysis is Sasha Disko's historical description of the 'multiple masculinities' of German male motorcycle riders.[49]

The early history of automobilism is quite different from that of cycling, although we should emphasize the *continuity*, rather than the breach, between the two movements.[50] Sydney Aronson enumerated ten ways in which (American) cycling culture showed a fundamental continuity with car culture: resistance from horsemen and pedestrians, assembly-line techniques in production, the emergence of repair shops, club culture, its attempted exclusion of fast-riding 'scorchers,' legislation and regulation, revival of wayside inns through peri-urban and inter-urban wandering, and the prominent role of women.[51] Mary Kennard's personal development from a cyclist into a motorist stands for many who made the same move; Leblanc started to write in *L'Auto* in 1903. There was also continuity institutionally, because of the cultural and political dominance of automobile clubs (gathered transnationally in the Association Internationale des Automobile-Clubs Reconnus, AIACR, 1904). They were soon joined by cycling clubs that in many countries started to accept motorists as well, after their car tourism had become a massive phenomenon.

But the crucial difference with cycling culture was the car's competing three non-animal propulsion systems, each thriving in their own cultural context, the technical genealogy of which also goes back, like that of the bicycle, to deep in the nineteenth century. Experimental steam-engine-powered buses appeared on British streets in the 1820s, followed by a second wave of steam-driven passenger cars half a century later. Experiments with electric batteries also go back to the early nineteenth century; they were first seriously tried in trams (and some buses) in the 1880s in the US, until their unreliability prompted investors to opt for the battery-less feed by overhead wire during the 1890s. We should be aware that this was a truly transnational 'movement': proponents and opponents alike were kept precisely *au courant* of developments through dozens of trade journals, sport magazines, and club publications. In Europe, the first serious experiments with passenger cars were electric (in Paris in the early 1880s), well before, according to the current popular belief, 'the car' was 'invented' by Carl Benz and Gottlieb Daimler in 1886 (or, if you are a French chauvinist, by Édouard Delamare-Deboutteville three years earlier).

At this point it is worthwhile to ask ourselves how best to tell the story of the 'victory' of the internal combustion engine or petrol car over its two rivals without falling into the trap of finalism, determinism, or teleology that the founders of SCOT warned us against, let alone avoiding national rivalries in popular myth formation. One possible approach (that enables us at the same time to finish the methodological argumentation that we began when we constructed the bicycle narrative) is to take the popular idea of the technical 'superiority' of the gasoline car at face value (an idea even to be found among scholars questioning the 'feminine' character of electric propulsion, as we will see later) and investigate early car experiments from that perspective.

There is no doubt, then, that internationally all experts, engineers, and entrepreneurs of horse taxi fleets alike opted for electric traction as the 'superior' technology when they started to contemplate fleet modernization around 1900, often after extensive experiments or thorough study of all three options (steam, electricity, petrol). Fleet tests with gas cars simply did not take place, because of their technical unreliability, as one fleet director in London told a journalist. In 1897, this director sent so-called Bersey cabs onto the streets, specially constructed on the basis of a mix of horse carriage and tram car technology, only to be taken off the streets two years later. The London specialized press blamed bad management as well as technology (defective lead-acid batteries, expensive and delicate tires) for this failure. A similar fate occurred when the largest horse taxi fleet company in Paris decided to electrify in order to prepare for the world exhibition of 1900, after it had set up a special training circuit for its drivers. In the US, hundreds of so-called hansom cabs (the driver at the back, above the passengers) were produced and applied from 1997 in New York (300 cabs in 1901), Boston (250), Washington (100), and some other cities, equipped with an innovative semi-automatic system of battery exchange, for instance in a station on Broadway, where the batteries could be charged during the night and fully charged ones were exchanged for empty ones whenever the need arose. The collapse of this initiative at the turn of the new century was directly caused by 'the greatest banking scandal of the decade,' which kept from the public eye that here too, technical deficiencies were to blame, again caused by battery plates disintegrating due to the shocks from the solid tires that replaced the puncture-prone and very expensive pneumatics. It is these cars that distorted the famous census of 1900, suggesting that they were dominating the car market in the US. When correcting for such utilitarian applications, however, the census clearly indicated that the steam car in private possession was the most popular (contrary to the later popular belief that electric propulsion dominated), a conclusion supported by anecdotal evidence from the trade press.

The idea of a defective electric vehicle technology (heavy batteries, short range, expensive tires), still popular today among some pockets of electric vehicle skeptics, is based on this first generation. This idea neglects the performance of the second generation of only half a decade later, successful not so much in car design (the new taxis kept their 'antique' appearance, reminiscent of the horse-drawn age), but in new, sturdier types of batteries (with an energy density [an indication of their quality] comparable to the ones still in use during the electric vehicle waves in the 1970s and 1980s) and tires. Such second generation taxicabs were introduced in Germany, and from there in the Dutch cities of Amsterdam and Haarlem, based on a French car design (Krieger). In Amsterdam a fleet of forty to eighty Kriegers functioned to the satisfaction of its founders until the early 1920s. Its fleet management could be studied in detail because the maintenance reports by the leading engineer have survived: they convincingly show that, had people at the time wished to 'go electric,' this would technically have been fully justified. The success of the Ataxes, as they were called after the company name, was due to a clever management innovation that made technical deficiency no longer crucial: in the charging station a specialist mechanic replaced defective battery plates and tires for new ones against a subscription fee of some cents per driven kilometer. Municipal governments in the Netherlands and Germany supported the electrics against gasoline alternatives, because of their polluting exhaust smoke and dripping petrol, which threatened to dissolve the asphalt pavement when waiting at the taxi stands. Electrification of the fire engine fleet in several German cities with the same batteries from the largest national manufacturer, AFA, resulted in the construction of more than 200 electric trucks in 1914, including several hybrid (petrol-electric, but also even steam-electric) versions, competing effectively (because of their possibility of an immediate start) with the steam propulsion system considered by many experts the ideal propulsion for heavy vehicles. These fleet experiences with vehicles weighing more than 6,000 kg, together with comparable experiences with thousands of delivery trucks in the US, showed that electric propulsion was able to perform successfully also in the domain considered particularly suitable for steam power, through a combination of improved battery and tire technologies, and clever management. In the US, electric traction acquired forceful backing from many local electric utility companies, who not only deployed their own electric car and especially electric truck fleets, supported by charging stations, but often allowed private persons to use these stations for charging their vehicles as well. The rivalry between electric and gasoline proponents was fierce: in 1899 a French electric broke the 100 km/h threshold under a lot of international fanfare. The shipping of early electrics to the queen of Spain or a sultan in India was widely advertised. It seems that especially medical doctors preferred electric traction, probably because of their familiarity with electric equipment, which for others smacked too much of a (scientific) 'laboratory.' The US also saw a modest 'electric tourism' emerge, enabling more planned car journeys within a network of charging stations on the East coast, so much so that even Henry Ford announced in 1914 that he would develop a cheap electric with a range of 160 km, together with his friend Thomas Edison, who had previously developed a special sturdy battery for this purpose.

By 1909, when the American electric vehicle fleet had reached 20,000 units, annual production amounted to 3,639 electric passenger cars. The same year saw gasoline car production rise to 82,000 units. In 1910 American motorists enabled the first outburst of mass car diffusion; by then the size of the American car fleet had reached half a million. Whereas the Atlantic world in 1900 counted 15,000 motor vehicles (one-third in the US), at the outbreak of World War I more than two million were counted, 65 percent in the US, mostly built by Henry Ford, and nearly all driven by internal combustion engines.

If the gas car 'won' the battle of the propulsion systems despite its technical unreliability, we must assume that this was a virtue, rather than a drawback. This will not surprise us after our previous observation of the third, functional part of cycling adventure: (preparedness for) tinkering formed an integral part of its culture. Given the continuity between cycling and automotive culture, it will neither surprise us that, when *Daily Mail* owner Alfred Harmsworth granted his first (steam-driven) car its silence, he added: "but so is a corpse."[52] The British motorist T. Chambers in 1907, writing about the technique of driving an electric, remarked that "there is not much sport in driving [it]. It is far too simple and too unexciting to be attractive. The fascination of the petrol engine to the man who is born with an engineering instinct is largely due to its imperfections and its eccentricities," and he predicted that if and when "the motor-car shall run with monotonous certainty, the main attraction of driving will have departed, and the amateur will turn its attention to ballons or airships, seeking for further difficulties to overcome."[53] And that is what happened from, say, the end of the first decade of the new century, although many millions, during the following decades, would still find ways to experience the automotive adventure in its tripartite form (speeding, roaming, tinkering) despite the airplane having taken over the position of ultimate adventure machine.

Virginia Scharff pointed at the hypocrisy of the masculine car culture that stereotyped the electric as a women's car in an effort to protect their culture from being 'invaded.' She is right, as we will see later in this chapter, but it is hard to deny that women initially *did* use the electric more than men, for instance in the flat city of Chicago with its many charging stations. They did not do so because the electric was an 'inferior technology,' an idea promoted by the gas car industry, echoed by Scharff, who pointed at "their manifestly lower power, frequently higher prices, and smaller range."[54] Recently, a detailed analysis of car ownership statistics and car advertisements makes it plausible that next to women, also some less adventurous men belonged to its early users, deploying it for their commute, a correction to existing scholarship in the same vein as occurred with early cycling history.[55]

The international breakthrough of the internal combustion car until World War I can be described in three phases of the dialectic between technical development and its respective 'relevant social groups' (in the terminology of SCOT), and of the cars' sequence of properties and functions (in the terminology of the 'dual nature of technology'). The first phase (until 1902) was characterized by a multitude of technical solutions, based on an amalgam, a 'bricolage' of bicycle and horse-carriage technologies. The substitution of the horse was not a linear process: horse breeding too came under the spell of modernity, resulting in stronger, faster animals considered to be urban "machines" rather than "fellow animals," and their number kept increasing, in some industrializing countries until well into the Interbellum.[56] The usual linear story is that the gas car was invented in Germany and developed into a sellable product in France, whereas the UK as the richest and most industrialized country, which within Europe had dominated the bicycle industry, was a latecomer in the car's initial breakthrough: in the rat race between capitalist states, Britain neglected to restructure its industry "by bolstering the ability of its empire to earn foreign exchange through the export of primary commodities."[57]

A less linear story needs to include both users and non-users. The automotive user cohort was a rich, aristocratic, and bourgeois avant-garde, who continued the group culture of the cycling years, but now mostly under the technical guidance of a chauffeur or mechanic. Although it is difficult to prove in the current state of the art of historical research, this author's impression is that the first users were single men, whereas during the second phase (until 1907) much heavier cars were developed and used for (large, bourgeois) family outings, driven by the man of the house. These intermediate five years are also the phase in which resistance against cars in the countryside (and in the US even in some cities) was revived, from mayors restricting access and driving speeds in their towns, or mobs (in New York) 'sabotaging cars by stoning them,' to a ban on automobiles in the Swiss canton of Grison (Graubünden) until the mid-1920s. The sabotaging and stone throwing formed a part of a class struggle with a typical socially marginal, anarchist aspect (perpetrators were called 'hooligans' and even 'apaches' in the motoring press), in some extreme cases ending in the beheading of motorists with ropes across the road. Remarkably, the 'anti-car crusade' subsided within a couple of years, when a general acceptance seemed to settle in, especially after the car had showed to be of societal use, for instance in the Netherlands during a mail strike in 1903 when members of the automobile club used their cars to deliver mail, protected by police and the military, or when the car played a similar role of assistance to the authorities after the earthquake in San Francisco in 1906. What also may have helped was that national governments acknowledged the existence of the movement through their legislation and regulation, quite quickly in fact, enabled not only because of the close class relationship between motorists and government officials, but also because of the complaints from another, more critical part of the ruling elite that became aware of the vehicle's dangers. This process was deepened after 1908, when the transnational organization PIARC (Permanent International Association of Road Congresses) generated a new breed of expert engineer-advisors to the national governments. Upon initiative by the French government, LIAT, AIACR, and PIARC cooperated to stimulate national states to place warning signs for dangerous road stretches and agree on a transnational traffic convention (1909) that more or less cemented the dominance of motorists over their rivals (cyclists, horse riders): in Paris, during the first PIARC conference of 1908, the 'road problem' (dust, dangerous curves, bad pavement width and quality) had been defined as a "car problem."[58]

The third phase begins with the Panic of 1907, an economic crisis that hit the French car industry especially hard, prompting its manufacturers to diversify toward utilitarian vehicles (trucks, buses, taxis; this was the moment that the electric taxicabs experienced competition from gasoline cabs that used the same garage-maintenance support 'invented' by electric cab fleet managers). But it was also a safety crisis, alerting the public to the car's dangers (in Berlin in 1907, a quarter of the registered cars had been involved in accidents), as well as a crisis of car technology (this was the moment that both the lead-acid battery and the pneumatic tire were fundamentally redesigned to allow for the ever-increasing vehicle speeds). In the latter case, we see a mechanism appear that seems to govern the competition of rival technologies in general.

[[Called 'inter-artifactual technology transfer,' the phenomenon was first observed in the social realm, and formulated by Herbert Marcuse as 'repressive tolerance,' defined as the inclusion of alternative activities in order to protect the hegemonic culture. Applied to the history of technology, the mechanism benefits the dominating technology, which plunders as it were the advantages of the technologies that threaten this dominance. For example, the sturdier tire was originally developed to compensate for the heavier electric car with its lead-acid batteries, but soon also used by petrol car manufacturers to enable higher speeds without too many punctures. Similarly, the electric motor of the gas car's rival was incorporated in the latter to avoid the risky starting of the engine by hand-crank, whereas its engineers were also incited, as it were, to develop a more silent engine to mimic the relative silence of the electric and the steam car. In the end, after World War I, even the closed body of the electric, and its character as a 'city car' with its fine-tuned speed regulation and its relative silence, were captured by the gas car paradigm. In the terminology of the 'dual nature of technology', not so much the technologies per se (the properties) were taken over, but rather the 'functions'; these were recreated by developing the 'properties' such that they fitted in the receiving artifact.[59]]]

Car *ownership* during this period remained an urban phenomenon with extra-urban *use*. The clubs tried to emphasize the utilitarian sides of what the national presses continued to call 'pleasure cars' (German statistics showed that the share of such cars was much higher, by about six to four, than that of commercially deployed cars). Only in the second half of the decade did American farmers start to motorize (European rural motorization only started in earnest after World War I), but they too indulged in holiday trips to the national parks (the first, Mount Rainier, opened in 1908), spending the night in 'autocamps' (from 1915). Or they took part in the 'See America First' movement, coinciding with the promotion of the countrywide Lincoln Highway. Living in the small towns, these farmers formed the first impetus to the country's massive motorization wave shortly before and during the war, at a moment when the motoring press started to signal the first forms of utilitarian use. A similar development in Europe was blocked by the war.

The car that emerged by the end of the first decade of the new century was a miniature edition of the very heavy family cars of the previous phase, although in some countries, especially in the UK, a more bicycle-type car was also revived from the first phase and gained quite some popularity. But it was the Ford Model T (and its more expensive European clones) that, from 1908/9, started to motorize a new type of user, called in the motoring press the 'Average Man' or the 'Small Capitalist.'

Transport historian Christopher Wells has argued that the breakthrough of the American gasoline car cannot be explained fully by its adventurous 'script.' He pointed at the role of small cars such as the runabouts and high-wheelers that had (similarly to the cyclecars in the UK, but on a much more numerous scale) a more utilitarian function. For him, the Model T's success must then be explained by acknowledging a kind of cross-fertilization between these utilitarian tendencies and the genius of Henry Ford to create a (high-wheeled) miniature version of the European Mercedes-as-adventure machine.[60] I agree: the existence of a minority subculture of 'domesticated' car use must be acknowledged, if only to contrast this with hegemonic adventurousness.

Although from the mid-1900s long-range touring started both in Europe and the US, including continental crossings by women, and by the end of the decade such crossings were undertaken by entire families, practicing such journeys still required a tripartite adventure, including the functional one: when William Vanderbilt in 1914 undertook a winter trip through France and Italy, he had a puncture every 750 km. It was in this phase, then, that an entire maintenance, repair, and fuel delivery infrastructure was built up, spearheaded by the Ford Company with its 'preventive maintenance,' a very lucrative way to allow a new breed of self-driving motorists to roam the countryside.

In his recent dissertation, the British historian Craig Horner has formulated a plea to reconsider the class base of the motorist of this period. Basing his argumentation on the same sources as we used when discussing the class base of the cycling craze and after, Horner added a study of Cheshire vehicle registrations between 1904 and 1914 to the debate, observing how motor vehicles "exchanged hands many times over just a few years, ending up in a working-class address in a suburb." Dubbing the phenomenon "modest motoring," he pointed at the cyclecar and similar extra-light vehicles, as well as the motorcycle, the latter in many countries dominating the registration statistics until well into the 1920s, in impoverished Germany even to the end of the Interbellum.[61]

There exists, indeed, evidence that in the US, where the motorcycle did not dominate the market like in Europe and where it became associated with a "deviant,"[62] highly adventurous minority, secondhand cars were "snapped up eagerly by the farmers of the Middle West" from the moment they started to buy cars in large amounts from 1910, a year in which 20,000 of the 150,000 new buyers bought low-priced 'runabouts.'[63] In some European countries, the sudden surge in motorcycle registrations (for instance in the French department of Cher between 1908 and 1909) also indicates that 1907/8 constitutes a transnational turning point in the appreciation of motorized mobility. In Germany, membership of the largest national automobile club ADAC more than doubled after 1909 from 13,000 to 28,000 within five years. By then, even the Catholic Pope gave automobilism his blessing in his *Benedicio vehiculi seu currus*.[64] But there does not exist historical evidence that the transnational car culture before World War I expanded beyond a prosperous middle class in any significant numbers. Even in the UK, it may have been the 'clerks' that co-constructed bicycle culture, but not significantly the culture of automobilism.

So, even if the automotive culture in some countries trickled down into some lower-middle-class secondhand car and two-wheeler pockets, it remained dominated by pleasurable, leisurely use, many aspects reminiscent of what had been developed within cycling culture.[65] For a deeper understanding of this culture we need not only to study the written accounts of car journeys but also (and especially for those experiences that were considered to be 'unspeakable') fictional renderings.

A detailed study of the dozens and dozens of travel accounts in the Dutch club journals, for instance, enabled us to construct a picture of a culture in which the 'cyborg' topos (the lust for 'unity' between body and machine) and the neurasthenia theme (the mostly masculine illness of nervousness for which car driving—like cycle riding previously—functioned as a therapy) appeared as a matter of routine, embedded in impressionist neologisms to describe the multisensory experiences of the surrounding landscape. By the beginning of the 1920s such travelogues had become so formulaic and boring that one journal editor invited literary figures to contribute.

This editor could draw from a rich literary tradition that we will now investigate more precisely than we were able to do for the bicycle craze.

[[For an analysis of such what we call 'autopoetic' sources (including sources that do not deal explicitly with automobilism, but contain 'autopoetic' scenes) we need some tools. The term 'autopoetic' is derived from literary analyst Lawrence Buell's use of the term 'environment-poetics.'[66] The most

> important tools stem from the insight that next to the obvious *content* level, we also (and more so, considering existing reservations among a recent generation of social scientists about the 'representation' of fictional sources) need to go two levels deeper, to the *symbolic* level and, still deeper, to the level of what I have called, borrowing from urban planner Marc Desportes's historical dissertation, the *affinity* between the practice of writing (or film-making) and driving.[67] We have used this concept already in our analysis of cycle culture, but it can also be observed much earlier, for instance in Wolfgang Schivelbusch's famous description of the experience of train travel as 'panoramatic,' a style of vision 'rhyming,' so to speak, with the panorama as a new visual urban form of entertainment.[68]

Charles Grivel called the car (from a literary point of view) a 'writing machine' (*Schreibmaschine*, literally a typing machine) or (from a filmic perspective) a 'vision machine,' adding that the role of the car in film (and by extension in literature) is "not primarily thematic."[69]

The obvious starting point for our analysis is a Belgian-French group of literati, who literally practiced 'affinity' by simultaneously pioneering automobilism and a new style of writing, a group so far not included as such in the world-literary canon.[70] Prominent among them was Maurice Leblanc, who became world-famous because of his novels on Arsène Lupin, the gentleman-thief and detective, and was asked by the *New York Times* to report on a spectacular armed siege by the police of the anarchist, car thief, and robber Jules Bonnot in 1912 in a Parisian suburb, while thirty thousand Parisians looked on. The invitation was no coincidence: the relationship of cars and violence was a popular theme among this group. Leblanc's Belgian friend and brother-in-law Henry Kistemaeckers wrote a short story (published in 1904) in which a car owner ties his wife to the grill of his car, driving into death, 'drunk on speed, angered by hatred.' *Les annales politiques et littéraires* in 1907 printed contributions of the group members, in which Leblanc praised the car as a tool to solve a situation in a tragic manner, and Octave Mirbeau (rich and red) called the car more dear and useful than his library, thus prefiguring the plebeian violence of Filippo Tommaso Marinetti and his fellow Futurists, who often figure in the popular perception as the true automotive iconoclasts and who may be remembered because of Marinetti's open hostility toward the bicycle in his Futurist Manifesto. Together with the German Expressionists during World War I, they were the first to consistently establish the car as an urban vehicle. Mirbeau's aristocratic, self-mocking attitude in *La 628-E8* (1907, partially pre-published in *L'Auto* and *L'Illustration*) is perhaps the most violent and aggressive, not only in the depiction of a trip through Belgium, the

Netherlands, and Germany in 1905, but also in the destruction of his linear narrative into a collection of impressionist tableaux. These writer-motorists read the 'script' of the car as a celebration of male violence and aggression, misogyny and condescension to the petty bourgeois, just at the moment that the latter started to become interested in the car. This is not to deny that some group members (such as Nobel Prize winner Maurice Maeterlinck, or the Belgian Cyriel Buysse) abstained from violence and aggression, but they all mobilized the symbols of the tripartite automotive adventure (time, space, function: speeding, roaming, tinkering) in some way or another, anthropomorphizing their car as an actor forcing them to speed up. In this they formed a part of a travel writing tradition in which the car was used as protective cocoon, for instance in American Paul Konody's *Through the Alps to the Apennines* (1911) or in *L'Espagne en Auto: Impressions de voyage* (1906) by the Belgian Eugène Demolder, during his southern voyage throwing coins at the Spanish Other. In the next chapter we will call this behavior 'ironic,' and will speak of the 'ironic car' as a quintessential element in the history of automobilism.

No wonder, then, that the 'adventure machine' as characterized above was celebrated in popular romance and youth literature. Alice Williamson's series of novels from 1902 blur the distinction between fiction and travel account, based as they are on her and her husband's travels, in which the 'conquest' of the woman is thematized, an adventurous masculine skill of seduction and car driving at the same time, the woman looking in admiration to the male driver. Similarly the American Motor Boys series (written by Clarence Young, 1900–17) or Katherine Stoles's Motor Maids series (1911–17) and many more have cars, ships, submarines, electric hydroplanes, and even an electric runabout (*The Speediest Car on the Road*, as the subtitle ran of one sequence of the Tom Swift series from 1910) drive the narrative's adventures. The first silent movies celebrated automotive adventurousness as well, more violently even than our written sources, for instance in Cecil Hepworth's *Explosions of a Motor Car* (in which motorists' limbs fly through the air) and *How It Feels to Be Run Over* (both 1900). One-fifth of the 500 pre-World War I American movies depicted road crashes, including, for instance, Thomas Edison's *Happy Crashes* (1909), from which the 'victims' escape unscathed.

Novels by the mainstream literati from this period, no less informative about the mood of contemporary automobilism than popular culture, give a more mixed picture. Marcel Proust and Nobel Prize winner, active car driver, and 'colonial adventurer' Rudyard Kipling play with the creative power of the driver, depicting him (in the latter's short story in his *Traffics and Discoveries* [1904] and in the former's "Impressions de route en automobile" [Road Impressions in a Car, published in *Le Figaro* in 1907]) as a musician or an organ player, describing car driving as a haptic, multisensorial experience, much more than the usual characterization of car driving as a visual experience. Indeed, John Urry's famous 'tourist gaze' applies more to the railway experience than to that of the early cars: car drivers seem only to be able to throw a furtive 'glance' at objects in the landscape.[71] According to literary scholars Proust added to the catalogue of automotive adventure the narcissistic experience of 'inversion,' in which, during his car trip through Normandy, the outer world seemed to move around the car, the static driver in the middle, but here too we must conclude that continuity reigns: Schivelbusch already observed inversion in the train traveler.[72]

Other mainstream writers are more ecstatic about the driving experience, most of all the iconic German Otto Julius Bierbaum, famous because of his travel account from 1903 in a borrowed car, but fantasizing about the car as a devilish vehicle in *Das höllische Automobil* (1906, The Car from Hell), or the British Marie Corelli's 'green' motoring novel from 1901, in which she not only criticizes the 'he-females' on bicycles, but also depicts the car as a flying demon. It seems that car critique among literati was most outspoken in the UK, such as in E. M. Forster's *Howards End* (1910), which continues H. G. Wells's rejection of the British upper- and middle-class culture, or in the Scot Kenneth Grahame's satirical *The Wind in the Willows* (1908), which exposes the car's 'seduction' to drive ever faster in an utter explosion of narcissism.

There were counter-voices, however, perhaps most particularly represented by lesbian feminist and New Woman Vernon Lee, who as one of the few novelists and other commentators tried to formulate a female adventure in her plea for an empathic 'reasonable travel' without harassing others, in her essay "The Motor-Car and the Genius of Places" (1905). And, of course, there is the other iconic Anglophone Edith Wharton, less inclined, like Vernon Lee and other women writers, to include violence and open aggression in her novel-travelogues, and dealing with the fictional vehicle (including its driver) in a special way by hardly mentioning it (or them), a rhetorical device that became very popular among literati after World War II, when the car had become so ubiquitous that it was always there, but never mentioned, as an always present 'absent car' (see Chapter 4). Thomas Mann, too, depicts the car primarily as a vehicle of status and power (in *Königliche Hoheit* [Royal Highness], 1909), rather than extra-urban adventure. The American autopoetic universe of this period also seems to be less aggressive and violent, perhaps best represented by Theodor Dreiser's book-length account of a car trip in 1915 (*A Hoosier Holiday*, 1916), although the country's movie culture suggests otherwise, as we saw: a separation of adventurousness along class lines.

The belletristic utterances mentioned here are only a part of an investigated sample of 150 novels, travelogues, and films made by fifty artists from seven countries. Taken together, their *contents* are dominated by four thematic clusters: the pride of developing technical skills; the experience of flight as an expression of orgiastic feelings of mobility in an immobile body; nervous restlessness ushering in the craving of high speed (engendering powerful fantasies of independence and personal sovereignty, especially in women); and anti-urban landscape experiences (including experiences of the colonial Other as well as travel into the past) resulting from furtive glances rather than a gaze; this all larded with violence and aggression so ubiquitous that it seems synonymous with driving per se. Speed as such was well known by now (trains reached higher speeds than cars), so it was speed *change* or acceleration (in curves) and deceleration (through braking) as well as personal regulation of speed that thrilled car occupants through their haptic, bodily responses, linking this speed and speed change with danger.

At the *symbolic level* we may characterize the car as a 'complex signifier,' around which a symbolic web was spun, consisting of the car as monster, as a woman to be tamed, as a protective shell or cocoon (despite the car's open body in this period), but also as a grenade.

At the level of the *structural analogies* between writing (and filming) and driving, we conclude that artists were not so much better observers than their less talented motoring fellows, but they were experts of experience, trained in translating (verbalizing, visualizing) their experiences. They stood for an increasing cohort of less well-equipped motorists, readers of the club journals mimicking the literati in their travel accounts sent to the journals. Together, they constructed a corporeal, eroticized, expansionist (imperialistic), highly exciting experience, in space, in time, and in technology, heavily gendered as an aggressive, violent flight in and of a shell. Many motorists, and especially the literarily inclined ones, described their experiences as transcendental, based on a multisensorial feeling of power toward everything and everyone outside of the automotive capsule, a hyper-experience that placed the driver outside of the individual Self, either toward a group feeling for others, or, more abstract, a feeling of a 'communal identity.'[73] There were counter-voices (women in particular), but automotive masculinity dominated.

There were counter-voices indeed, for the first time researched in earnest by American historian Virginia Scharff who, in a reaction to second-wave feminism's focus on the differences between men and women, emphasized that car-pioneering women, "invad[ing] a male domain," were equal to men in their skills and their adventurousness. "Women's part in creating the car culture" consisted primarily in overcoming the prejudice that pleasure while motoring is "worth getting unless it holds excitement or danger." Scharff quoted the first American female race car driver, Joan Cuneo, who complained that manufacturers failed to build cars "especially adapted to women's use," and she points at the electric starter as an example (of 'inter-artifactual technology transfer,' I would add) of the incorporation of what many considered "feminine frills" (but reinterpreted by Scharff as covert male desires) into "standard automotive equipment."[74]

In the wake of Scharff's study, Australian historian Georgine Clarsen not only unearthed a car factory run by women in Scotland during the war, she also analyzed the travelogues and other autopoetic texts by eleven Anglophone women car drivers, showing "the extra work that women needed to perform to produce themselves as competent motorists," including "[t]rain[ing] your ear to distinguish the slightest sound foreign to the consistent running of the engine," a faculty that made women especially suited for this pastime, according to Cuneo.[75] As a very early example of this (fictional) technical sovereignty we have Mrs. Mary Kennard, whom we encountered earlier as women's advisor in cycling matters, whose autopoetic novel *The Motor Maniac* (1902) depicts a female protagonist who indulges in the functional adventure of tinkering and maintenance. "She had read that it required a trained electrician to repair an induction coil if wrong," but she and her driver manage to repair it themselves, continuing their "very happy day" by speeding, at 45 km/h, as fast as an "express train." She has no time for gazing at the landscape, "her eyes glued to the road ahead."[76] Likewise, Florence Luscomb and Magaret Foley's 'suffrage auto tour' between 1911 and 1915 in the US was no less adventurous, including a car race with an opponent in the elections. Scharff's and Clarsen's approaches must be seen as a response to those positions that argue that autopoetic travelogues were mostly written by men because "the essence of adventure lies in taking risks and exploring the unknown."[77] Such controversies alert us to the need for mobility historians interested in car-cultural history to spend energy to study its gender relations, including resolving the lack of studies into the homosocial and homoerotic bonds between men and their cars, the latter a fully ignored aspect of the history of automobilism.

And yet, there is also another part of the automotive adventure that seems to escape not only both Scharff's and Clarsen's efforts, but also academic mobility history in general. It lives on mostly in amateur histories, where it functions as a kind of exotism, or in the rare work of an academic especially interested in the social aspects of folklore. As we saw during our previous dealing with cycle culture, certain members revived the 'gymkhana' skills of earlier horse dressage and carriage handling days, a tradition continued in the flower courses and *concours d'élégance* of car culture. From the "gala lawn fêtes" in chic American Newport, where Mrs. O. Belmont had set up an "obstacle course," driving her imported European car decorated with "yellow field flowers," via the "Longchamps fleuri" in no less chic Spa, Belgium, followed by a lantern parade watched by an international crowd of fifteen thousand, back to the "riding gymkhanas through bowling pins" of the 'deviant' American motorcyclists, such practices suggest a largely forgotten undercurrent of a culture focused on the commodity character of the car and the celebration of its status, as well as the inflow of more women in these cultures.[78] A British motorcycle gymkhana as late as 1914, the women clad in "many-coloured costumes," offered "ballon bursting from sidecars, ring tilting motorcycle and sidecar musical chairs, as well as non-motoring events such as hat trimming, pea-guessing, balloon racing and bun-biting," while other ones included a symbolic "tug of war" between motorcyclists and car owners.[79]

Eva Chen, whom we met when discussing the role of women in the fledgling cycling cultures, recently provided a convincing argument about the supportive role of women in the acceptance and breakthrough of automobilism in the West, based on their "aestheticizing" role of the movement, a practice she called an "active strategy," thus transforming the car "from a machine of adventure and athleticism into a commodity of conspicuous consumption." Although my analysis of the Interbellum car movement (see the next chapter) suggests an addition rather than a substitution of adventure and status, it cannot be denied that women motorists and passengers "actually helped to dispel the public's 'motorphobia' and enhance motoring's reputation because of their fashionable elegance." Supported by quotations from the British press, Chen discovered the beginning of the taming of the adventure machine, enabling the historiography of automobilism to integrate car culture with body design and fashion, in both aspects of the (wo)man–machine cyborg.[80] The adventure became comfortable, the adventure machine morphing into an adventurous fashion machine.

One other aspect of the motoring culture of the first period should be mentioned: its military character.[81] Especially in the institutional realms of the movement (the clubs, the contests and races) the influence of the military and the national armies was substantial. The German cultural historian of mobility Kurt Möser and the Swiss economic transport historian Christoph Merki have shown the intimate connection between automobile sports and the military culture of honor, manliness, disgust for the decadent bourgeois society, and contempt of death, and how preindustrial chivalrous traits pervaded early bicycle and car culture.[82] The result was a collectivist, defeminized, masculine culture of male bonding with built-in aggression and violence.

When World War I broke out in 1914, military logistic science had prepared for a 'railroad war.' Like the bicycle and the car, railway construction can be traced back to the early nineteenth century, but after the Austro-Prussian and Franco-Prussian wars of 1866 and 1870 a trebling of the European track length to 300,000 km took place, mainly realized in France with Germany's railway superiority of 1870 in mind. National armies also closely followed the rapid development of bicycle and car technology, however, resulting in the setting up of a Kraftfahr-Abteilung in 1907 in Germany and the purchase of some cars and trucks by the American Quartermaster Department in 1908. Around the same time, in several countries car club members gathered in Volunteer Motor Corps, driving military officers around during exercises and ready to have their cars confiscated in the case of war. Whoever still had some hope of an electric future would be disillusioned by the results of extensive tests by military experts, but their rejection may have had less to do with the usual argumentation (especially the electric's short range, compensated, as we saw, by expertly managed charging stations), but because of the vulnerability of these stations in case of war (a bomb hit would disable all vehicles in one blow) and perhaps also because of the feminine reputation of electric propulsion. The car's individualized culture was a military asset.

Popular historical memory celebrates the heroism of the 'taxis de la Marne' (September 1914); at closer look the confiscated Parisian taxis caused a logistic chaos of traffic jams and vehicle defects, but managed to bring spare parts and ten thousand soldiers to the front 50 km away, a media event that according to a recent study had a symbolic function but was "militarily insignificant."[83] Much more impressive, in fact, was the 'Voie Sacrée' (Sacred Road) to Verdun, a feat of military logistics that kept a slippery and icy 7 m wide and 75 km long road occupied by a conveyor belt of around 10,000 motor vehicles per day (trucks, buses, ambulances, passenger cars, motorcycles), one every fourteen seconds, transporting 90,000 men and 50,000 tons of material per week at an average speed of 15–20 km/h. 1200 workers spread 700,000 tons of limestone to repair the pavement. Drivers were forbidden to overtake or stop, while groups of mechanics repaired broken-down units or shoved them into the ditch. In eleven months, 2.4 million men and 2 million tons of freight were transported over the road, accomplished by a Direction des Services Automobiles (DSA) that commanded 13,000 cars and 31,000 trucks (many of them requisitioned) in 846 sections, in which 1,500 officers and 75,000 soldiers (7,200 mechanics among them) kept the fleet running.

By the end of the war the British army had 35,000 trucks in France. When America entered the war in November 1917 they bought their cars and trucks from French factories and subsumed their men under the French centralized approach. Despite the enormous presence of horse traction, in total the Allied powers deployed 170,000 motor vehicles behind the trench lines (including some British tanks, after 1916), against 130,000 on the German side.

The later popularity of the Marne taxis (versus the near-forgotten logistics of the DSA) may be related to the adventurous culture spun around the automobile during the prewar years. The auto club in the neutral Netherlands warned its members not to indulge in war tourism, because their cars could be confiscated by the military. Autopoetic texts from war participants showed the close relation between car driving and violence: 'We were intoxicated,' a contributor to the German magazine *Der Motorfahrer* (The Motorist) remembered, describing his car trip to the front 'over dead bodies . . . All full of blood, dripping from the wheels and mudguards.' The war enabled writers to reach adventurous sublime, for instance in Anton Fendrich's *Mit dem Auto an der Front* (1915, To the Front by Car) or Paul Grabein's *Im Auto durchs Feindesland* (1916, Through Enemy Land by Car). Famous writers such as Gertrude Stein, Ernest Hemingway (as a Red Cross volunteer), and even Edith Wharton indulged in 'thanatourism,' the latter, in her propagandist *Fighting France* (1915), throwing off her tranquil habit, discovering adventure: her car 'flew' and 'dashe[d]' over the roads. In the US, Willa Cather let her fictional farmer boys go to the war, where they enjoyed a 'tingling sense of ever-widening freedom' in her controversial novel *One of Ours* (1922).

No less adventurous were the 16,500 women that joined the American Expeditionary Force, or the 57,000 women in the British Women's Army Auxiliary Corps, or the 4,000 French women in the DSA, some of them driving ambulances. British drivers, "racing motorcars along foreign roads like adventurers exploring new lands," according to feminist historian Sandra Gilbert, needed a state driving license as well as a mechanic's license.[84] The Red Cross Motor Corps was a big logistic machine of 600 motor vehicles, more than half trucks, maintained and repaired in two large garages in Paris by skilled mechanics.

Summarizing, there can be no doubt that the pre-World War I car was predominantly seen and used as a violent, masculinized 'adventure machine.' Some scholars, such as Ian Boutle in his pioneering, but somewhat exceptionalist history of English racing, have expressed doubts whether "'man-machine' futurist dialogues" as mobilized by cultural historians such as Kurt Möser (and further developed by this author) are correct and should not be replaced by "English articulations of the ideal driver [which] owed more to familiar elitist visions of self-control."[85] It is true that the more idyllic literary expressions of car enthusiasm can be interpreted as representing a minority of 'modest' motorists. Even then, there can still be no doubt about mainstream literary utterances standing for hegemonic real-life violent adventurism, not as fantasy, but as 'real' feelings of unease. It is hard to speculate about the source of those more idyllic utterances, whether fictional or travelogue, in which the car was depicted in the midst of an individualized idyll in a natural wilderness, but some form of compensatory energy may have been involved here. Whatever the reason, however, it warns us not to take the car propaganda of the clubs and literati at face value, because even these 'modest motorists' caused accidents during and after the 'safety crisis of 1907' (see earlier) and wreaked havoc in the countryside, as the early anti-car resistance testifies. When the war ended, many thousands of young men had become 'streetwise' as drivers of trucks or of ordnance passenger cars chauffeuring officers, or had been educated as car mechanics ready to drive and repair the trucks and taxis during the postwar boom years. The war resulted in the first contours of a (militarized) 'car system.'

Notes

1. Michael Adas, *Machines as the Measure of Men: Science, Technology, and Ideologies of Western Dominance* (Ithaca, NY: Cornell University Press, 1989), 357. See for the following: Gijs Mom, *Globalizing Automobilism: Exuberance and the Emergence of Layered Mobility, 1900–1980* (New York: Berghahn, 2020), 59–71.
2. Daniel R. Headrick, *The Tools of Empire: Technology and European Imperialism in the Nineteenth Century* (New York: Oxford University Press, 1981), 3, 181–83, 96; on "strategic thinking": Ronald E. Robinson, "Conclusion: Railways and Informal Empire," in Clarence B. Davis and Kenneth E. Wilburn, Jr. (with Ronald E. Robinson) (eds.), *Railway Imperialism* (New York: Greenwood Press, 1991), 175–96, here 192.
3. Giovanni Arrighi, *The Long Twentieth Century: Money, Power, and the Origins of Our Times* (London: Verso, 1994), 62.
4. Aparajita Mukhopadhyay, *Imperial Technology and 'Native' Agency: A Social History of Railways in Colonial India 1850–1920* (London: Routledge, 2018), 4; Divall quoted in Aparajita Mukhopadhyay, "Transport Users, and Historie(s) of Transport: 'A View from Below,'" *Journal of Transport History* (2023): 1–15, here 4.
5. See for instance Marian Aguiar, *Tracking Modernity: India's Railways and the Culture of Mobility* (Minneapolis: University of Minnesota Press, 2011).

6. Dirk Hoerder, *Cultures in Contact: World Migration in the Second Millennium* (Durham, NC: Duke University Press, 2002), 419, 425, 436; 150: Sebastian Conrad and Klaus Mühlhahn, "Globale Mobilität und Nationalismus: Chinesische Migration und die Re-Territorialisierung des Nationalen um 1900," in Birgit Schäbler (ed.), *Area Studies und die Welt: Weltregionen und neue Globalgeschichte* (Vienna: Mandelbaum, 2007), 217–51, here 219.
7. Adam McKeown, "Global Migration, 1846–1940," *Journal of World History* 15, no. 2 (2004): 155–89, here 167, 169; last quote: Matthias Middell and Katja Naumann, "Global History and the Spatial Turn: From the Impact of Area Studies to the Study of Critical Junctures of Globalization," *Journal of Global History* 5 (2010): 149–70, here 155.
8. Yong Sang Lee, "Session Description" (Session "History of Korean Railroad: Geography, Geopolitics and Imaginative Geography", 2023 GMHC&T2M Annual Conference, Seoul, Konkuk University, 25–28 October 2023); Woo Hyun Park, "Directional Monopoly in Colonial Railroad Construction: The Longitudinal Route Centrism of the Joseon Railroad" (ibid.).
9. M. William Steele, "The Speedy Feet of the Nation: Bicycles and Everyday Mobility in Modern Japan," *Journal of Transport History* 31, no. 2 (December 2010): 182–209, here 185–86, 191.
10. David Arnold, "Subaltern Streets: India, 1870–1947," in Tariq Jazeel and Stephen Legg (eds.), *Subaltern Geographies* (Athens: University of Georgia Press, 2019), 36–57, here 36 ("neglect"), 42–44; see for the following also Mom, *Globalizing Automobilism*, 72-85.
11. Stewart Lone, "Japan and the Age of Speed: Urban Life and the Automobile 1925–1930" (discussion paper No. IS/05/494, London School of Economics and Political Science, July 2005), 1–13, here 2–3; Christopher Madeley, "Kaishinsa, DAT, Nissan and the British motor vehicle industry" (Ibid.) 14-38, here: 15, counts only four privately owned cars in Tokyo by 1906; Donald F. Davis, "Competition's Moment: The Jitney-Bus and Corporate Capitalism in the Canadian City, 1914–29," *Urban History Review* 18, no. 2 (October 1989): 101–22, here 103 (zenith). 'Jitney' seems to have been derived from Louisiana French Creole slang for a small coin: *jetnée*. Mia Bay, *Traveling Black: A Story of Race and Resistance* (Cambridge, MA: Belknap Press of Harvard University Press, 2021), 158.
12. Ibid., 151, 162. The quote is from the Little Rock *Gazette*.
13. James Francis Warren, *Rickshaw Coolie: A People's History of Singapore 1880–1940* (Singapore: Singapore University Press, 2003 [1986]).
14. Mom, *Globalizing Automobilism*, 120.
15. Hans Peter Hahn, "The Appropriation of Bicycles in West Africa: Pragmatic Approaches to Sustainability," *Transfers* 2, no. 2 (Summer 2012): 31–48, here 35; David Arnold and Erich DeWald, "Cycles of Empowerment? The Bicycle and Everyday Technology in Colonial India and Vietnam," *Comparative Studies in Society and History* 53, no. 4 (2011): 971–96, here 979, 982–83.
16. David Strand, *Rickshaw Beijing: City People and Politics in the 1920s* (Berkeley: University of California Press, 1989), 25–26.
17. Natsume Sōseki, "The Diary of a Bicycle Rider," in *Spring Miscellany and London Essays* (Boston: Tuttle, 2002) (first published in 1909 as *Eijitsu Shōhin*), 139, 147.
18. Natsume Sōseki, *The Wayfarer; Kōjin* (New York: Perigree, 1982 [1967]), 260.
19. See for the following Gijs Mom, *The Electric Vehicle: Technology and Expectations in the Automobile Age* (Baltimore: Johns Hopkins University Press, 2004),

32–35, for a more detailed reconstruction, but the current text discusses also more recent scholarship, justified through endnotes.

20. Theodore R. Schatzki, *Social Practices: A Wittgensteinian Approach to Human Activity and the Social* (Cambridge: Cambridge University Press, 1996); James J. Gibson, "The Theory of Affordances," in Robert Shaw and John Bransford (eds.), *Perceiving, Acting, and Knowing: Toward an Ecological Psychology* (Hillsdale, NJ: Lawrence Erlbaum, 1977), 67–82. See for the following Gijs Mom, *Atlantic Automobilism: Emergence and Persistence of the Car, 1895–1940* (New York: Berghahn, 2015), 35–36, and Gijs Mom, "The Dual Nature of Technology: Automotive Ignition Systems and the Evolution of the Car," in Christian Huck and Stefan Bauernschmidt (eds.), *Travelling Goods, Travelling Moods: Varieties of Cultural Appropriation (1850–1950)* (Frankfurt: Campus, 2012), 189–207.
21. Madeleine Akrich, "The De-Scription of Technical Objects," in W. E. Bijker and J. Law (eds.), *Shaping Technology, Building Society* (Cambridge, MA: MIT Press, 1992), 205–23.
22. Wiebe Eco Bijker, *The Social Construction of Technology* (Eijsden [PhD thesis University Twente], 1990).
23. See for the following: Gijs Mom, "New Mobility Studies: Niche to Mainstream; History and Social Sciences in the Mobility Turn" (*Transfers*, forthcoming).
24. Rüdiger Rabenstein, *Radsport und Gesellschaft: Ihre sozialgeschichtlichen Zusammenhänge in der Zeit von 1867 bis 1914* (Hildesheim: Wagenbach, 1991). Stephen Kern, *The Culture of Time and Space 1880–1918* (London: Harvard University Press, 1983), often cited by social science scholars, places the movement in the context of the period's longing for 'speed.'
25. Jehan de la Pédale, *Pédalons! Avec une préface de Richard O'Monroy (Pompon) et une lettre de Richard Lesclide* (Paris: Véloce-Sport, 1892), 17.
26. As related, for instance, in Catherine Bertho Lavenir, *La roue et le stylo: Comment nous sommes devenus touristes* (Paris: Editions Odile Jacob, 1999).
27. Yoonjoung Choi, "The Bi-Cycling Mr Hoopdriver: Counter-Sporting Victorian Reviving the Carnivalesque," *Critical Survey* 24, no. 1 (2012): 102–15.
28. Ibid., 103.
29. See, for instance, Una Brogan, "Cycling and Narrative Structure: H.G. Wells's *The Wheels of Chance* and Maurice Leblanc's *Voici des ailes*," in Marian Aguiar, Charlotte Mathieson, and Lynne Pearce (eds.), *Mobilities, Literature, Culture* (Cham: Palgrave Macmillan, 2019), 237–57.
30. Peter Bailey, "White Collars, Gray Lives? The Lower Middle Class Revisited," *Journal of British Studies* 38 (July 1999), 273–90, here passim.
31. Ibid., 275 ("status"); G. L. Anderson, "The Social Economy of Late-Victorian Clerks," in Geoffrey Crossick (ed.), *The Lower Middle Class in Britain 1870–1914* (London: Croom Helm, 1977), 113–33, here 113 (most numerous); John Lowerson, *Sport and the English Middle Classes 1870–1914* (Manchester: Manchester University Press, 1993), 8–15.
32. Ibid., 14–15.
33. According to Glen Norcliffe, *Critical Geographies of Cycling: History, Political Economy and Culture* (Farnham: Ashgate, 2015), 161, in the ordinary era of the 1880s club members and "unattached" cyclists were about equal in numbers.
34. Eva Chen, "Its Prohibitive Costs: The Bicycle, the New Woman and Conspicuous Display," *Journal of Language, Literature and Culture* 64, no. 1 (2017): 1–17, here 17n11 (schoolteacher); long quote: David Rubinstein, "Cycling in the 1890s," *Victorian Studies* 21, no. 1 (Autumn 1977), 47–71, here 59.

35. Ibid., 50–51, 56, 65.
36. J. W. Allen, *Wheel Magic or Revolutions of an Impressionist* (London: John Lane, n.d. [1909]), 36–37, 46–47.
37. Rüdiger Rabenstein, "The History of German Workers' Cycling Association, Solidarity," *Cycling History 11* (Proceedings, 11th International Cycling History Conference, 23–25 August 2000), 160–67.
38. Clare S. Simpson, "Capitalising on Curiosity: Women's Professional Cycle Racing in the Late-Nineteenth Century," in Dave Horton, Paul Rosen, and Peter Cox (eds.), *Cycling and Society* (Aldershot: Ashgate, 2011), 47–65, here 53.
39. Jean Orselli, *Usages et usagers de la route: Requiem pour un million de morts: 1860–2010* (Paris: L'Harmattan, 2011), 76; for the UK and the Netherlands see Mom, *Atlantic Automobilism*, 609, 614.
40. Phillip Gordon MacKintosh, "A Bourgeois Geography of Domestic Bicycling: Using Public Space Responsibly in Toronto and Niagara-on-the-Lake, 1890–1900," *Journal of Historical Sociology* 20, no. 1/2 (March/June 2007), 126–57, here 129, 136, 140, 145–48.
41. Una Brogan, "Domesticated Mobility: Exploring the Limits of Women's Liberation through Cycling in British Literature, Guide Books and Periodicals, 1890–1910," *Cycle History 29* (London, 2018), 55–60, here 57; Eva Chen, "Its Beauty, Danger, and Feverish Thrill: Speed and Cycling Women in Fin de Siècle Fiction," *MFS Modern Fiction Studies* 63, no. 4 (Winter 2017), 607–27, here 608 ("crucial"), 611 (racing).
42. Allen quoted in Chen, "Its Beauty," 615; collision: 616; Eva Chen, "Freedom or Constraint: The Speeding Bicycle and the Industrial Subject in H.G. Wells's *The Wheels of Chance*," *Victorian Literature and Culture* 45 (2017): 433–47, here 433 ("machine").
43. H. G. Wells, *The Wheels of Chance: A Holiday Adventure* (London/New York: J.M. Dent/Macmillan, 1896); speeding on 34, 40, 44, 79 (an average of 9 km/h), 98, 133 (chase), 136; objects on the road: 30, 34, 46, 48, 55, 86 (bad city road); dust: 96, 133; puncture: 51; inspection of the bicycle: 275; violence: 82 ("squirting"), 148; quest: 13. This reprint of the original publication emphasizes the adventure as well as leisure in its subtitle. In the 1896 British edition, Wells calls Hoopdriver's 'adventure' an "idyll" (on 142). Brogan worked with the American edition that has the subtitle: *A Bicycling Idyll*.
44. Patrick Parrinder, "Biographical Note," in H. G. Wells, *The History of Mr. Polly* (London: Penguin, 2005 [1910]), vii–xii, here x ("propagandist"); John Sutherland, "Introduction," in ibid., xiii–xxix, here xxiii (Fabian), xxix ("genteel"). Example of 'propagandism' in ibid., 122–23; hostility: Jeremy Withers, *The War of the Wheels: H.G. Wells and the Bicycle* (Syracuse, NY: Syracuse University Press, 2017), 154–56.
45. Choi, "The Bi-Cycling Mr Hoopdriver," 102 (recreational vs. sport); Wells in 1897: "the bicycle in its earlier phase . . . brings the adventurous side uppermost," quoted in Brogan, *The Alternative Modernity of the Bicycle*, 40.
46. Withers, *The War of the Wheels*, 17 (poor); Norcliffe, *Critical Geographies of Cycling*, 163 (bourgeois); Paul Smethurst, *The Bicycle: Towards a Global History* (New York: Palgrave Macmillan, 2015), 74 (1913 and 1930).
47. Ibid., 157.
48. R. W. Connell, *Masculinities* (Cambridge: Polity Press, 1995). See also David Morgan, "Class and Masculinity," in Michael S. Kimmel, Jeff Hearn, and R. W. Connell (eds.), *Handbook of Studies on Men & Masculinities* (Thousand Oaks, CA: Sage,

2005), 165–77; subordination quote in Toby L. Ditz, "The New Men's History and the Peculiar Absence of Gendered Power: Some Remedies from Early American Gender History," *Gender & History* 16, no. 1 (April 2004): 1–35, here 2 ("reinvented" on 6).

49. Sasha Disko, *The Devil's Wheels: Men and Motorcycling in the Weimar Republic* (Oxford: Berghahn, 2016).
50. See for the following Mom, *Atlantic Automobilism*, 62–63, and Mom, *The Electric Vehicle*, 6–13, 39–63, as well as Chapters 2 and 3 (64–201).
51. Sidney H. Aronson, "The Sociology of the Bicycle," *Social Forces* 30 (1951–52): 305–12.
52. Alfred C. Harmsworth, *Motors and Motor-Driving* (London, 1902), 45.
53. Quoted in Lord Montagu of Beaulieu and E. Wilson McComb, *Behind the Wheel: The Magic and Manners of Early Motoring* (New York, 1977), 177.
54. Virginia Scharff, *Taking the Wheel: Women and the Coming of the Motor Age* (New York: UNM Press, 1991), 49.
55. See Josef Taalbi, "On the gendering of the early American electric car," *Environmental Innovation and Societal Transitions* 54 (2025) 100934.
56. Joel A. Tarr and Clay McShane, "The Horse as an Urban Technology," *Journal of Urban Technology* 15, no. 1 (2008): 5–17, here 6.
57. Arrighi, *The Long Twentieth Century*, 278; richest: James Foreman-Peck, Sue Bowden, and Alan McKinley, *The British Motor Industry* (Manchester: Manchester University Press, 1995), 6.
58. Mom, *Atlantic Automobilism*, 576.
59. See Gijs Mom, "Conceptualising Technical Change: Alternative Interaction in the Evolution of the Automobile," in Helmuth Trischler and Stefan Zeilinger (eds.), *Tackling Transport* (London: Science Museum, 2003), 9–46.
60. Christopher W. Wells, "The Road to the Model T: Culture, Road Conditions, and Innovation at the Dawn of the American Motor Age," *Technology and Culture* 48, no. 3 (July 2007): 497–523, here 506, 512, 519, 522.
61. Craig Horner, *The Emergence of Bicycling and Automobility in Britain* (London: Bloomsbury Academic, 2021), 104.
62. Sammy Kent Brooks, "The Motorcycle in American Culture: From Conception to 1935," unpublished dissertation (Washington, DC: George Washington University, 1975), iv.
63. G. F. Carter, "Automobiles for Average Incomes," *Outing* 55, no. 4 (January 1910): 410–19, here 416.
64. Vittorio Marchis, "Not Cars Alone: An Italian Story," in *Motu Proprio: ACI's First Hundred Years, Heading Onward* (Rome: Automobile Club d'Italia, 2005), 87–94, here 89.
65. This has been confirmed by quite original research (using many personal diaries of British citizens) by Colin Pooley, Jean Turnbull, and Mags Adams, *A Mobile Century? Changes in Everyday Mobility in Britain in the Twentieth Century* (Aldershot: Ashgate, 2005).
66. Lawrence Buell, *The Future of Environmental Criticism: Environmental Crisis and Literary Imagination* (Malden, MA: Blackwell, 2005), 55.
67. Marc Desportes, *Paysages en mouvement: Transports et perception de l'espace XVIIIe–XXe siècle* (n.p. [Paris]: Gallimard, 2005), 60. Only after the completion of my analysis did I realize that the French sociologist of literature Lucien Goldman (one of the central theorists of my literary studies before I went into engineering and the history of technology) had observed a similar phenomenon that

he called 'structural homology.' See Gisèle Sapiro, *The Sociology of Literature* (Stanford, CA: Stanford University Press, 2023), 23–24.

68. Wolfgang Schivelbusch, *Geschichte der Eisenbahnreise: Zur Industrialisierung von Raum und Zeit im 19. Jahrhundert* (Munich, 1977), 59–62.
69. Charles Grivel, "Automobile et cinéma I," in Michel Bouvier, Michel Lartouce, and Lucie Roy (eds.), *Cinéma: Acte et présence* (Quebec: Éditions Nota Bene, 1999), 63–85, here 63, 68; Charles Grivel, "Automobil: Zur Genealogie subjektivischer Maschinen," in Martin Stingelin and Wolfgang Scherer (eds.), *HardWar/SoftWar: Krieg und Medien 1914 bis 1945* (Munich: Wilhelm Fink, 1991), 171–96, here 182.
70. See for the following Mom, *Atlantic Automobilism*, Chapter 2 (133–226).
71. Jonas Larsen, "Tourism Mobilities and the Travel Glance: Experiences of Being on the Move," *Scandinavian Journal of Hospitality and Tourism* 1, no. 2 (2001), 80–98; John Urry, *The Tourist Gaze: Leisure and Travel in Contemporary Societies* (London: Sage, 1990).
72. Schivelbusch, *Geschichte der Eisenbahnreise*, 59.
73. For more examples see Mom, *Atlantic Automobilism*, 401, 517, 649.
74. Scharff, *Taking the Wheel*, 13, 21, 29 (Cuneo quote), 66, 67.
75. Georgine Clarsen, *Eat My Dust: Early Women Motorists* (Baltimore: Johns Hopkins University Press, 2008), 22.
76. [Mary] Kennard, *The Motor Maniac: A Novel* (London: Hutchinson, 1902), 172, 190, 199.
77. Susan Bassnett, "Travel Writing and Gender," in Peter Hulme and Tim Youngs (eds.), *The Cambridge Companion to Travel Writing* (Cambridge: Cambridge University Press, 2002), 225–41, here 225; for the 'suffrage auto tour' see Tim Cresswell, *On the Move: Mobility in the Modern Western World* (New York: Routledge, 2006), 214.
78. William H. McGaughey, *American Automobile Album* (New York: E. P. Dutton, 1954), 14–15; Donald Weber, *De blijde intrede van de automobiel in België 1895–1940* (Ghent: Academia Press, 2010), 45; Brooks, "The Motorcycle in American Culture," 29, who makes an explicit connection between gymkhanas and "the entrance of women."
79. Joshua Richard Butt, "Aspects of the Automobile's Diffusion in the North-West of England 1896–1939," unpublished dissertation (Manchester: Manchester Metropolitan University, 2018), 70.
80. Eva Chen, "Pretty Women Don't Drive: Early Women Motorists and the Taming of the Motor Car," *Women's Studies* 49, no. 2 (2020): 149–79, here 151 ("aestheticizing"), 152 ("strategy"), 164 ("elegance").
81. See for the following Mom, *Atlantic Automobilism*, Chapter 3 (227–81).
82. Kurt Möser, *Geschichte des Autos* (Frankfurt: Campus, 2002); Christoph Maria Merki, *Der holprige Siegeszug des Automobils 1895–1930: Zur Motorisierung des Strassenverkehrs in Frankreich, Deutschland und der Schweiz* (Vienna: Böhlau, 2002).
83. Holger H. Herwig, *The Marne, 1914: The Opening of World War I and the Battle That Changed the World* (New York: Random House, 2009), 262.
84. Quoted in Schraff, *Taking the Wheel*, 93.
85. Ian Boutle, "'Speed Lies in the Lap of the English': Motor Records, Masculinity, and the Nation, 1907–14," *Twentieth Century British History* 23, no. 4 (2012): 449–72, here 453–54.

CHAPTER 2

Migratory and Other Subversive Mobilities, 1920–1940

When feminist writer Yosano Akiko (pen name of Shō Hō) in the second half of the 1920s traveled through Manchuria upon invitation by the railway company, she not only used the train, like Natsumo Sōseki did in the previous period (see Chapter 1), but was also driven in an automobile "rac[ing] along the road at a fearful breakneck speed."[1] She could do so because Japan, next to its railway imperialism, also undertook a massive campaign of 'development through road building': 30,000 km of (largely unpaved) roads were laid out, followed in the late 1930s by plans to join Manchuria, Korea, China, and Indo-China through a Great East Asian Highway. In 1942 even the construction of an expressway between Dalian and Harbin was started, straight as an arrow, inspired by the autostradas in fascist Italy. Roadbuilding, we will observe in this book, is often done ahead of motorization: whereas in mainland Japan around 1930 about 120,000 motorized vehicles were registered (a modest 56,000 of them passenger cars, the rest motorcycles, buses, and trucks), Manchuria counted only 3,000 as late as 1936. But Yosano also used the sedan chair during her trip, and walked, an activity apparently worthy of being mentioned emphatically.

After the Great Kantō earthquake of 1923 (which destroyed the rickshaw factory), the government imported Ford T-trucks and adjusted them to passenger transport to replace the electric trams. A rare study of Japan's early road motorization emphasized its *utilitarian approach*, not only expressed in the use of passenger cars as semipublic buses (as we saw in Chapter 1), but also by focusing, in the industry's advertising, on the "nuts-and-bolts level; that is, value for money" and on "low fuel consumption."[2] Whereas during Japan's "economic miracle" of the 1930s passenger cars were produced in the three assemblage factories of Ford, General Motors, and Chrysler, domestic manufacturers started to produce trucks and buses. The latter acquired some fame because of the female conductors ("bus girls," *basu gāru*), until they were replaced by "one-man buses" in the 1960s.[3] Interbellum road traffic in Japanese cities was "extremely" unsafe (especially for pedestrians), a "traffic hell" to which authorities responded with "large-scale safety campaigns" from 1921.[4] Such campaigns were no doubt also aimed at the cyclists, whose number had increased much more than that of cars, from four million in 1925 to eight million in 1938, their spread forming a near-perfect logistic diffusion curve. The bicycle also spread to the countryside, and gradually shifted its societal meaning from "the speedy feet of the working class" to a vehicle for (middle-class) women, the opposite of what at the same time was happening in several countries in Western Europe and the US, where the bicycle became considered a 'poor man's vehicle,' as we saw in the previous chapter.[5]

According to world system theorists, Japan may be considered a "semi-periphery," as it belonged to "a middle stratum [of countries]—both agent of and subject to exploitation."[6] In this chapter we will investigate how the 'car society' (as a successor to the 'car system' of the previous period) was constructed within three different mobility cultures of cores and (semi)peripheries, which at the same time helps us reconsider this dichotomy, opening the possibility of identifying a third geopolitical entity, the 'Global East,' which we will do in Chapter 4.

In the case of China it is less well known that Sun Yat-sen, when he became president of the new republic after the revolution of 1911, not only designed a railway network of 200,000 km, but also envisioned constructing "one million miles of road," as well as the production of a cheap car. By the time of his death in 1925, however (at a moment when China counted 8,510 cars), only 8,000 km were finished, on which sinologist Frank Dikötter discovered the existence of a fledgling motor bus network; by the 1930s ten body-making factories existed in Beijing alone. Yet recent research by sinologist Thorben Pelzer revealed that by 1930 30,000 km of roads were constructed by Sino-American civil engineers, often educated in the US, built by hundreds of thousands of Chinese (including schoolgirls) under "coercive and punitive work regimes," the funding (for the "relief work" alone amounting to a staggering $50 million between 1921 and 1935) coming not only from the Red Cross but also from (clearly other kinds of) "Chinese people." The roads were thus built way ahead of demand, and apparently exclusively for an elite (in Guizhou province, the road had to cater to one registered car), because the use of donkeys and some types of carts was forbidden. When famines subsided in the 1930s, this kind of roadbuilding stopped.[7] Yet the layeredness of China's mobility configuration was dominated by human-powered river transport. Whether human-powered or by bus, tourism was possible in "a region encompassing the eighteen Chinese provinces of the former Qing empire" (Tibet, Xinjiang, and Mongolia were out of reach).[8]

From Sun's death onward, China's development followed three distinct paths. The China International Famine Relief Commission and its (mostly American) missionaries were followed by advisors from the League of Nations (founded after the "European war" of 1914–18), who managed to help construct, before the worldwide Depression of the early 1930s, perhaps up to 56,000 km of roads for China's 40,000 to 50,000 motorized vehicles. Second, after the defeat of workers' movements in some Chinese cities in 1927, from 1934 a Communist "migration of a nation" on foot and horseback was set in motion, 86,000 men (and 35 women), a Long March of six thousand miles over "atrocious or nonexistent roads," attacked by Italian bomber planes and 'developing' the rural population through the confiscation of land. And third: less well known are the development campaigns of some warlords, such as Yan Xishan in Shanxi province west of Beijing, who, impressed by Soviet developments, built roads with assistance from the American Red Cross, set up truck and bus services, used these roads to fight against his neighboring warlords, and sabotaged the nationalist government's anti-Japan activities. Cars in this phase were exclusively owned by the very top of the ruling and the urban middle class, from governors, ambassadors, and Shanghai gangsters to Russian military advisers and members of the corrupt government. As late as 1934, the army possessed only 1500 'vehicles.'

In urban traffic flows, these cars were easily overshadowed by carriers of sedan chairs, cyclists, pedestrians, and rickshaw pullers, who, in Beijing in October 1929 for instance, initiated violent 'wild riots' against the emergence of the motorized bus and the tram. Their working conditions, worsened after a new wave of migration from the countryside during the Depression, were increasingly seen as 'inhumane' by middle-class reformers and literati of the May Fourth Movement, one of them being Lao She (pen name of Shu Qingshun), whose novel *Luotuo Xiangzi* (Camel Xiangzi, serialized in 1936 and 1937, translated into English as *Rickshaw*) gathered an iconic status as it drew a harsh picture of the life of a rickshaw puller, his family, and the community of pullers.[9]

Protests against the inhumane conditions of rickshaw culture spread to several Asian cities with large rickshaw cultures (such as British Malaya and Dutch Palembang), but reform efforts (for instance in China's Guangdong province, after a puller insurrection in 1927) came to nothing. Knowledgeable observers saw the middle-class indignation as ignorance about a form of life (rather than a mere profession) of the very poor, one of the few opportunities for migrants from the countryside (where their workloads were often much harsher). However, the Guomindang government announced the abolition of hand-pulled rickshaws after the war with Japan, a measure implemented by the Communists when they took over the country in 1949. The banning of 'informal' transport would be a continuous phenomenon during the entire twentieth century and beyond, irrespective of the political inclination of the governing elite.

Whereas China's mobility culture during the Interbellum rested on its inland navigation system, India could boast of a fully functional railway system. Mahatma Gandhi used it for his 'triumphant train tours of 1920' after he had become the leader of the Indian National Congress. Both countries had what Western observers would call an 'underdeveloped' road network, India because the colonial government's transport policy aimed at protecting its railway interests, resulting in an extreme layeredness of its road mobility, dominated by thousands of bullock carts, but without many cars, trucks, or buses. What characterized Indian road mobility in this period were the various types of peddlers, discovered by subaltern studies scholars. These belonged, for instance, to the gypsy tribes of the Banjaras or the wandering castes of Bhojpuri singers, whom the government classified as 'criminal tribes' because of their nomadism, in some cases going back more than seven centuries. Several of them became the first truck drivers in the 1920s and 1930s.

⟦ *Subaltern Studies* was founded—as a publication—by Ranajit Guha in 1982 in Britain to write an Indian 'history from below' to counteract the usual elitist history of Indian nationalism and the independence movement, using the remarks on the subaltern by Antonio Gramsci. It was also a reaction to the British Marxist historical elite (and its analyses based on 'class'), especially when Partha Chatterjee (related to Maoist student organizations) and others introduced Foucaultian and other poststructuralist critiques of orthodox Marxism, as well as literary criticism and postcolonial theories, which made the movement very influential in American academic circles, and beyond. One of its strong methodological aspects was its efforts to write thick descriptions of "counterinsurgency" by Indian peasants through a form of reading "against the grain," thus giving agency to those who produced cultures (and mobilities) who have no voice in historical records or move against the grain of hegemonic mobilities.[10] As we saw in Chapter 1, subaltern studies scholars neglected the history of the urban subaltern, the areas where we locate, in this book, the emanations of subversive "subaltern mobilit[ies]," a term coined by subaltern studies scholar David Arnold.[11] ⟧

Whereas India can be seen as an example of a strict application of what in the metropole was called "coordination" (the struggle between rail and road), Malaysia's mobility history during the Interbellum can serve as an example of the limits of the alleged universalism of Western concepts and methods. Anthropologist Thomas Williamson described the construction by the British colonizers of roads, first to the mining centers and rubber plantations, then covering the entire country, the British colonial government providing loans to officials "to buy a car or motor cycle" to drive over "3,000 miles of excellent public roads" where they could indulge in the automotive adventure "no different from Europe."[12] Here too, a large, racialized variety of road users could be found: Malays rode horse carriages and motor vehicles, Chinese controlled the garage business and were bus and tram conductors and rickshaw pullers, whereas the trams were driven by Sikhs. Not a single European was involved in Malaya's transport business.

In the Dutch colony of Indonesia, too, roads were first constructed toward the coal mines, then expanded to a network of 44,000 km, on which not only a diverse array of draught animals was to be seen (camels, donkeys, buffaloes, llamas, and elephants), but also the *dokar* (pony cart). By the beginning of the 1940s, shortly before Japan would invade the colony, 50,000 cars were to be found (mostly locally assembled in a General Motors factory, from 1927), causing a similar 'coordination crisis' as in the West. In the colonies especially, the Western middle class's feeling of insecurity led to an emphasis on the conspicuousness of their car consumption: status seemed to acquire as great a value as adventure.

Whoever is interested in the emergence of automobilism under colonial conditions should study Africa, if only because during the last fifty or so years historically interested anthropologists have provided a solid empirical foundation for such a study. As in other continents, mobility history in Africa is first and foremost a history of migration, which started well before the colonial period: for African migrants a possible dichotomy between a mobile West and an immobile Rest is an absurd construction indeed.

Already in 1982, world historian and economist Patrick Manning's very detailed study of Dahomey (current-day Benin) analyzed the struggle, around the turn of the twentieth century, for hegemony between two 'societies': caravans, head porterage, and sailing ships in a mercantile setting versus capitalist mobilities in the form of steam ships and steam trains. Against this background the usefulness of roads must be seen "as an effort in socialization rather than investment."[13] Other studies also emphasized the utilitarian (as opposed to adventurous) trade flows occupying the roads, and an aggressive 'coordination' policy to protect the government's railroad interests. During the entire period the railways in Africa remained the basic support of colonial exploitation, and when motorized road transport started to increase, a trucking monopoly was either given to the railways (Tanzania) or the road was forced into a feeder role to the railways (Ghana). Barrel rolling, head porterage, palanquin and machilla carrying, ox wagons, and sleds (or used canoes) in sandy areas, as well as cycling, formed the mobility substrate on which car and truck mobility was superimposed. Very rarely, Sahara crossings by car or crossings of the entire continent, from the Cape to Cairo, could be seen, amply described, in their quadruple adventurousness (speeding, roaming, tinkering, conquering), in travelogues for an eager public in the Global North.[14] Diana Strickland, Lady Dorothy Mills, famous literati such as Graham Greene, Evelyn Waugh, André Gide, they all undertook motorized excursions into the unknown, the latter accompanied by sixty porters and thirty-five cases sent ahead in two vans. These travelers followed in the wake of the highly publicized 'cruises' by car manufacturers between 1926 and 1932 into Africa and China, supported by French intelligence services, and in the case of the crossing of the Sahara opening the desert for winter tourism by car and bus.

As to the more mundane forms of motorized transport in this period: Jan-Bart Gewald is one of the few Africanists who studied early automobilism (in Zambia).[15] He confirmed the pivotal role of the 1920s, which puts this history in synchronicity with the Asian histories. What seems especially striking is the importance of the 'ironic' character of the car in the case of colonial automobilism: local Western elites lived in an automotive bubble, and during their 'administrative tours' their cars distanced them from the environment, protecting them from the Other, just like linguistic irony distanced the speaker from her audience.

⟦ The literary technique of irony is defined in its verbal form as "the use of words to express something other than and especially the opposite of the literal meaning," a form of nonliteral language use (next to, for instance, the use of metaphors), which has helped the (middle-class) writer keep his distance to the world, just like the car cocoon brings the (middle-class) motorist close to the gazed-at other, but always in a distanced way, protected by the shell.[16] Borrowing from literary and linguistic studies of irony, we can define the car as a vehicle of nonverbal irony, but in a sense other than the "Louvre's glass pyramids," proposed by Rebecca Clift as an "ironic play on our prototypical knowledge of pyramids as solid." Rather, the car's irony rests on the affinity between the production of the car trip and the production of texts, both resulting in a distancing effect toward the world. Modern studies of linguistic irony distinguish between "wolves" (those who understand irony) and "sheep" (those who do not).[17] As we will see in this and the coming chapters, irony is racialized and class-based. ⟧

Recently, this picture of Africa's fledgling motorization during the Interbellum has been enriched considerably by Andrew Denning's analysis from a continental, "transimperial" perspective, emphasizing the 'Latin' flavor of automotive imperialism in Africa, led by the French and Italians, supported by the Belgians and the Portuguese. They constructed the bulk of the somewhat more than half a million kilometers of roads, used by almost 300,000 motor vehicles by the end of the Interbellum, a project Denning calls "a failure," mostly because of its destructive (instead of constructive) character, pointing at a calculation by economic historian Elise Huillery who showed that "French West Africa (AOF) subsidized France more than France subsidized AOF . . . The investment in African roads was one of African labor, not European capital." Only if one adds the 140,000 km in South Africa, with its very different history, does the British impact on the continent's mobility infrastructure appear significant.[18]

For British Tanganyika (Tanzania) STS professor Joshua Grace added a very welcome native perspective to this history, claiming that a history thus told invites us to "explore and reflect on alternatives to the better known forms of automobility."[19] Arguing somewhat iconoclastically that the 'mobility turn' "translates poorly to historical experiences in East Africa," he reads alternative narratives basically as narratives of migration. This already becomes clear when tracing the etymology of the Kiswahili term for automobile: *gari* is Hindi for "vehicle," brought to Tanzania by migrants from South Asia, where they traveled on their bullock and donkey carts along the trade routes. Tanzanians, Grace explains, were "walking to the car," literally so, if we consider the history of "long-distance African walking." Sometimes they walked in front of it, as in the case of the colonial official who decided that he had to be pulled during his inspection tour of 80 km in his car, the gearbox in neutral.[20]

Next to the colonial and semi-colonial mobilities of Asia and Africa, Latin American semi-peripheral mobilities knew their own specificities. One of these no doubt was the desire for a Pan-American Railway and a Pan-American Highway, the latter a project initiated and mostly run by US experts who had the South identified as an 'obvious field for development.'[21] This became all the more important when, in the 1930s, the US car market began to show signs of saturation.

We have only a patchwork of information about the early motorization of the Latin American continent. Mexico's road network construction from below has been well covered by studies that show how the traditional form of voluntary labor (the *faena*) already before the revolution of 1910 had been mobilized to build public works. The first breakthrough of Mexican automobilism seems to be related to its revolutionary use ("beyond the confines of the urban elite") during the 1910s, when "rebel factions that had sprung up in previous years quickly clashed with the federal government and amongst themselves for control of the nation's transportation and communications infrastructure, a battle that had the effect of undermining much of the old railway and telegraphic network."[22] In the 1920s and 1930s, stimulated by the promise of North American long-range car tourism, and after the nationalization of the oil industry, local and national roads and even parts of the Pan-American Highway were constructed, on which a regulated truck sector was thriving. In 1940 Mexico counted 145,000 cars and 45,000 trucks.

Such campaigns aggravated the 'coordination crisis.' The economic historian Ricardo Salvatore observed that the railways (built mostly by the British) "had not solved the transportation problem of sparsely populated areas." International development specialist Fiona Wilson, for instance, showed how in the Peruvian Andes indigenous and immigrant "daredevils and adventurers," carrying goods and passengers, successfully competed with the British-owned railroad to the capital Lima.[23]

Early Latin American car enthusiasm is easily underestimated. Car densities in Argentina in the 1920s were comparable to those of France and Great Britain and those of Chile were comparable to the somewhat lower values of Germany and Italy. Ford and General Motors had set up assembly plants in the 1920s, but some 'differential commonalities' (Arjun Appadurai) with Europe, such as the late start of the regular automobile shows in Brazil (in 1923), suggest the lack of a prosperous middle class as a possible explanation (Brazil's car densities were comparable with those of Serbia and Bulgaria). Yet, despite this gap (or perhaps because of it, as the early Brazilian car culture could now be dominated by the wealthy aristocracy), the Automóvel Clube do Brasil's passion for racing was legendary, whereas in the capital Rio de Janeiro foreign observers were awed by the 'expensive high-power cars' in that city.

Environmental historian Shawn Miller, who wrote a book-length study of Rio's "transformation from a meaningful place for sociality, commerce, and leisure to a space of darting and daily death," did not mention these adventurous tropes.[24] Although he acknowledged that true mobile 'fluidity' started with the "rapid movement" of the car, his emphasis is on the urban "street" rather than the "road," which not only enabled him to dedicate an entire chapter to housing, the 'dwellings' of mobility studies, but also to observe especially the "high social status" character of the cars, which captured street space, each occupying "as much ground as a standing elephant." There were about 10,000 cars in 1926, one-tenth official cars. These were mostly driven by chauffeurs, Italian immigrants and men of African descent. And as if to support the Comaroffs' thesis about the periphery being the laboratory of the core, we read about early one-way streets "that astonished European visitors." By the 1930s this automotive culture displayed more attention to topics such as congestion and parking space than the precarious safety conditions of Rio's streets.[25]

Summarizing, in the so-called non-Western Interbellum we observe a remarkable synchronicity of motorization, roadbuilding, and protection of railway interests. Wherever a middle class or a colonial elite existed, local cultures of adventurous car use emerged, their differences related to the road mobility substrates that existed before the coming of the car. Mobility layeredness was universal, but differed per region. One wonders, then, what types of cars enabled these cultures: how and where were they designed and produced, and how far did they differ from the cars of the previous period? To answer these questions, we have to move back to the West.

The emergence of the 'Western' car as a socio-technological system can be described as a classic case of co-construction, of manufacturers, their engineers, and the potential customers (and several other stakeholders, such as the government).[26] The result of this process, which can be followed for the US in detail by close reading of the journal and conference transactions of the Society of Automotive Engineers (SAE), was the 'affordable family car,' which was defined partly in opposition to (but also in transnational cooperation with) European manufacturers. The process was one of 'translation' of the perceived customer wishes, which ended in the wish of the engineers and manufacturers to 'educate the public' and to distance the latter from the former.

What is less well recognized in the historiography is the active role of the potential 'users,' for instance in the costly process of transforming the car from an open 'shell' into a closed 'capsule.' During the war American users had added soft 'tops' and 'emergency curtains' to their open T-Fords to enable all-weather and all-season use, provoking a remarkably fast 'encapsulation' of the occupants by the manufacturers. Industry marketeers identified women as the new customers, whose alleged desires for 'comfort' made Virginia Sharff, as we saw (see Chapter 1), look at their masculine partners and husbands as perhaps the real drivers of this development. Nonetheless, in the 1930s Californian car dealers reported that one third to one half of their customers were women.

The case of the tires is even more convincing than that of the closed body: customers started to deflate their tires for less vibrations, against the advice of the experts, who developed special 'balloon' tires with more spring function. But then customers again started to deflate these tires, forcing manufacturers to develop not only super-balloon tires, but to redesign the entire wheel suspension, because the softer tires caused the front wheels to 'shimmy,' vibrating heavily around their vertical axis at certain speeds, thus jeopardizing the safe handling of the car.

When, in 1927, the attraction of Ford's Model T had subsided and General Motors took the lead in American car manufacturing, the commodification of car consumption ushered in an annual model change of a car that was carefully designed such that it could function at the same time as a city car but also, and especially, as a vehicle to indulge in speedy long-range tourism, weekend trips, and other forms of car leisure, enabled by the construction of a network of new highways. The urban characteristics of the new car seemed to rob the electric alternative definitively of its *raison d'être*: electric propulsion from now on (until the post-World War II years) was mostly to be found in industrial vehicles (such as forklift trucks) or commercial vehicles (such as the British 'milk floats'), or in very well-managed urban fleets such as that of the German post office.[27] Apart from adventure, the car now could also provide status, as a possession that created differences, such as depicted in novels by F. Scott Fitzgerald (*The Great Gatsby*, 1925) and Michael Arlen (*The Green Hat*, 1924).

Urban use, however, should not jeopardize its high-speed adventurous character. American motorists, the journal *American Motorist* observed in 1928, came in the grip of speedy 'go-fever.' 'Comfort' for this type of use was not only created in the form of a redesign of the gear box (so-called constant-mesh gears, making gear shifting easier), more powerful engines running on knock-free fuel (better resistant to the higher pressures inside the cylinders), and drive trains (adding so-called overdrives, enabling the engine to run on lower speeds, making less noise and thus functioning more economically), but also through 'softer' wheel suspensions developed by a new type of 'riding and body engineers.' Engineers and users together thus co-constructed a 'comfortable adventure machine.'

'Comfort' also implied the experience of domesticated danger. During the first three PIARC conferences until the beginning of the 1920s, there was still some doubt among experts over whether some safety technologies (such as lighting) were a problem of car manufacturers (forcing them to apply headlights and rear lights) or of road engineers and the state (lighting the road). But the power to define what a 'car' really 'was' was increasingly taken into the hands of the manufacturers: 'co-construction' from now on took place very indirectly. Whereas in 1921 72 percent of automotive patents in the US were still filed by individuals, by the end of the 1930s they were mostly filed by corporations.

Manufacturers hired university laboratories to test body vibrations and 'orchestrate' sound spectra such that a 'multisensorial room on wheels' was created, dominated by vision but crucially supported by a 'comfortable' haptic experience.

European manufacturers went through similar exercises, profiting from a reversal of knowledge flows over the Atlantic, more and more in the direction of Europe, but with one important difference: different layouts of roads and streets and different tax regimes (and perhaps also a different relation to the car adventure) made European engineers and manufacturers design stiffer suspensions, higher-revving engines, compacter cars, and in general a more 'sporting' handling of these cars.[28] Whereas Europeans drove their cars like 'courageous speed-hounds,' Americans were 'lazy drivers.'

The spread of this new car-in-the-making during the Interbellum over a new layer of consumers differed per continent, even if their common synchronicity was never affected. To take the UK as an example: the rise of middle-class income, the lowering of the price of cars, and a willingness to buy enabled an explosion of car ownership, from a hundred thousand in 1918 to two million in 1938, with by far the most used for "pleasure" (commuting accounting for 10 percent at the most), difficult to combine with Sean O'Connell's conclusion that the interwar years saw the shift from the car as luxury to one of "necessity."[29]

This synchronicity between countries could be maintained because less prosperous European would-be consumers motorized through the motorcycle. This not only reinforced the 'adventurous' character of the mobility culture in Europe, but it also introduced "two distinct classes of owners" of motorized vehicles.[30] The most extreme case was Germany: whereas the smoothly growing performance of German cars (in passenger-kilometers) was interrupted by the Depression, and thereafter received only a boost through special tax measures by the Nazi government and the promise of a people's car in mass ownership, not realized before the war (*Volkswagen*), the motorcycle's performance curve over the two decades did not show any sign of stagnation. The German *Arbeiter-Radfahrerbund* allowed motorcyclists to become members in 1928.[31] Also in the UK a thriving motorcycle culture can be observed during the Interbellum, perhaps also driven by "a new type of lower-middle-class worker—more secure, more 'modern' and more actively interested in politics. This was the better educated specialist such as the accountant, the draftsman, and the technician," entertaining "a new form of lower class bohemianism."[32] Whereas the car industry at the end of the 1930s calculated that upkeep of a car necessitated a minimum income that formed the upper threshold of an average worker's salary (a 'middle-class' car cost the same as a house), the motorcycle by that time was generally considered as "being dominated by working class riders." Apart from some utilitarian applications (for goods delivery and even as a "sidecar taxi"), commuting only figured modestly in its use profile (workers generally lived close to work and walked or used bicycles, of which there were fourteen million by the mid-1930s, double that of 1931), whereas the main application was for "pleasure" (such as holidays or weekend trips). During this period the UK lost its position as the world's largest exporter of motorcycles to Germany. Whereas in 1924 about one million light motor vehicles were counted on British roads (half of them motorcycles), by 1930 this had doubled (with only one-third of them motorcycles, an estimated 25,000 [3.4 percent] owned and driven by women); when at the end of the interwar years 2.5 million vehicles were registered, the absolute amount of motorcycles had nearly halved. In his dissertation, Christopher Potter analyzed several contemporary trade and club journals to find that clerks were the most common motorcycle owners, "being of an independent, educated, and adventurous nature." They needed to have a "willingness to tinker."[33] A similar analysis of journals for the Netherlands resulted in a comparable picture of pleasure driving with a more than proportional attention in the motorcycle press for women riders.[34]

Whereas in many European countries the motorcycle culture formed the adventurous motorization start-up for the lower middle classes, such a 'luxury' did not exist for citizens of the Soviet Union. The country became known for its 'roadlessness' (*bezdorozh'e*) after the Bolshevik revolution of 1917 (except for the highways connecting the large cities in the European part, built before the revolution). Avtodor (the Society for the Promotion of Automobilism and Road Improvement, founded in 1927, publishing its own journal *Za rulem* [Behind the Wheel]) and its president, the Left Communist Valerian Osinskii, deployed an array of activities to promote automobilism, most of all through a series of 'adventurous' car rallies, starting in the early 1920s, when most of the about 18,000 (mostly prerevolutionary) motor vehicles were to be found in Moscow and Petrograd, including the "Black Marias" or "Ravens" used by the secret service "on nightly roundups." Like in Western Europe, the travelogues in *Za rulem* were cast as adventure stories. When Joseph Stalin launched agricultural collectivization and national industrialization in 1928, Osinskii traveled to the Ford factories in the US, resulting in the construction of the Gor'kii Automobile Factory (GAZ), which started production in 1932, amounting to 130,000 vehicles in 1937. In 1929 Stalin had publicly dreamt of the USSR becoming "a country of metal, an automobilized country, a tractorized country" that soon would make his country "advanced" instead of "backward." In the absence of a middle class, in a society with a stratification of party members versus all the rest of the population, the newly produced cars were mostly distributed among the party elite. Voluntary roadwork doubled at the beginning of the 1930s when collective farms needed trucks (in 1938 a total of 630,000 could be counted), and Moscow even saw the beginning of a sixteen-lane highway construction, reminiscent of the 'mixed roads' originally proposed (but soon rejected because of the high costs) within PIARC before the war, but now built by convicts within the Gulag system. Shortly before the war the secret service (responsible for road maintenance) developed a plan for the construction of a modest road network of 3,700 km, centered in Moscow.[35]

At first sight different from all the countries dealt with up to now, Soviet mobility history during the Interbellum was characterized by 'socialist urban planning' emphasizing electric tramway networks, but, under the influence of Western planning (its handbooks translated into Russian during the early 1930s), Soviet planners debated their replacement by bus networks, something that only took place after World War II (see Chapter 3).[36]

In the US average household budget, car purchases overtook furniture in 1924. There, the first massive motorization wave started in the small towns shortly before (as we saw in Chapter 1) but especially during the war, when farmers' income increased considerably (and the price of the Model T decreased likewise). It then started to include the urban middle class, so much so that by the end of the 1920s three-quarters of American households possessed a car. The car thus helped urbanize the countryside, while at the same time enabling the creation of new types of suburbs (without public transport), and perhaps even triggering the Depression when the domestic market witnessed its first signs of saturation. Apart from the issue of scale (which of course should not be ignored), both continents showed synchronous developments: market saturation was also observed in the Netherlands, the UK, and even Germany. Saturation is a moving target: it expands with the change of preferences and income toward new groups and classes.

In the US this explosion of car purchase and use was not only the result of a more comfortable technology and an urge toward long-range tourism, however. As the car still refused to be reliable enough for a lay user, manufacturers, starting with the Model T, now began to organize 'preventive service' (after-sales service), whereas at the same time, in an astonishingly short period during the 1920s, the prewar pharmacy with its 'motor spirit' was substituted by a countrywide fuel-supply network of fuel 'stations.' In Europe, where similar developments took place, a fuel crisis and an agricultural crisis led to the forced addition of alcohols.[37]

[[In the collective co-construction of this story by two or three generations of mobility and transport historians, several myths have crept into the 'master narrative.' The most important ones are the 'toy-to-tool' thesis and the 'US-as-model' thesis. Whereas for the latter thesis we have already on several occasions pointed at the 'synchronicity' between European and American developments (and will continue to do so in the following chapters), for a refutation of the former thesis an analysis of the cries by the car lobby about the 'car as necessity' is asked for. The necessity argument pops up among historical actors as well as historians. Hitler himself called the car a 'utilitarian object' (*Gebrauchsobjekt*). And Virginia Scharff referred to the American 'car suburbs' to give credence to this thesis.[38] But several arguments can be brought against this.]]

The most important argument against the 'US-as-model' thesis is the evidence of a vast (and very probably dominant in the extra-urban car use profile) touring and tourism culture, both in the USA and in Europe. Contemporary car proponents also referred to the increasing urban use of the car, but here too there is ample evidence that commuting by car often took place (if at all) from the suburbs to the nearest train station, the husband brought there by his partner. Even in the US, workers generally could not afford to commute by car: in California as late as 1935, 80 percent of working-class families used the streetcar for this purpose (and, to place the car 'explosion' somewhat better in its historical context, 44 percent of all American families had no access to a car at all). The same applies to the UK: the pounding on the 'necessity' argument can therefore best be seen as part of a legitimation effort by a sector that precisely at the beginning of the 1920s came under attack because of its increasing unsafety (especially among urban pedestrians, often women and children, a condition described by a Californian newspaper as a class war), which triggered campaigns by the industry with parades and brochures, as we will see later.

And yet, when *Negro World* in 1924 called the car a "necessity rather than a luxury" for Black motorists, something quite different was at stake. Although middle-class Blacks also indulged in long-range tourism, historian Mia Bay has recently shown how "deeply racialized" American automobilism was: Blacks often had to stuff their cars full of food and other amenities and even petrol when they could not find a roadside restaurant, a motel, a fuel station, or an auto-camp catering to them, even if they used *The Negro Motorist Green Book* (from 1936). And when they entered the South they had to be careful not to offend white drivers by giving right-of-way or avoiding passing them. Even sharecroppers and poor Blacks felt the 'necessity' to buy a car (secondhand "rattletraps," reviving a rather 'necessary' form of the functional adventure of DIY repair), their car often being their only possession. The introduction in 1908/9 and the subsequent cheapening of the Model T gave many of them the opportunity to buy a new one (just as Ford allowed Blacks to be among the first to work in his factories), thus motorizing the Great Migration. A government overview from 1941 claimed that between 25 and 50 percent of Black and white sharecroppers owned a car, two-thirds of these secondhand. Wealthy Blacks, however, practiced a "prosperity . . . on parade" by purchasing "extravagant vehicle[s]" and cruising in them through the main streets of the big cities. Gretchen Sorin explained this by referring to the difficulties for Blacks to purchase houses. She warned against stereotyping Black car culture: although their precarious travel culture urged them to buy bigger cars, the iconic relationship between Blacks and the Cadillac is controversial. Although only 1 percent of Black car owners purchased a Cadillac, Black car ownership between the wars must have been substantial: by the end of the 1920s W. E. B. Du Bois found the "Jim Crow cars [of the railways] usually empty."[39]

Historian Sean O'Connell, for the UK, added a cultural rather than economic argument to the question of why more workers did not buy cars, despite an extensive secondhand market, the possibility of car sharing (which some British workers *did* practice), and the availability of purchase on credit: many of them considered the car a typical middle-class vehicle used predominantly for leisure and pleasure: this shows that economic arguments to 'explain' the spread of the car are not sufficient.[40] Detailed research by social and historical geographer Colin Pooley and social historian Jean Turnbull of hundreds of British diaries showed that in the UK during the 1930s only one-tenth of journeys to work were done by car.[41]

Now that we have seen emerge the car as machine and the classes of its buyers, the next question is which car culture was shaped by these two 'actors.' For the US it has been well established that the entire countryside was 'in a perpetual state of motion' during the Interbellum, and especially during the 1930s. There was first the migration to the suburbs, from within the city and from beyond. Then there was the blue-collar (and partially Black) Great Migration from the South to the northern cities (the flight from the Great Plains because of the drought alone involved 60,000 families). Third, in the early 1930s half of all cars in the country were driven to the National Parks and the auto-camps in the South and West; among them were the "transcontinentalists" who sped their cars up to cross the entire continent along the hardly paved Lincoln Highway. They numbered 20,000 in 1921 alone. These tourists, described by Warren Belasco as early as 1979 in a still unsurpassed study, encountered counter-flows of southerners looking for jobs and farmers prospecting for better land.[42] On a more regional and local scale, Sunday outings and evening spins were undertaken. Overall, average trip lengths increased from 125 miles in 1916 to 400 in 1936, but such trips (compared to other trip types, such as shopping and social visits) were most negatively affected by the Depression.

Similar developments can be observed in Europe. At a much smaller scale, road surveys in the Netherlands showed that on workdays half of the passing motorists practiced tourism, but in the weekends 'recreative traffic' dominated. German tourism under the Nazi regime was dominated, however, by collective 'package tours' by bus and especially trains, tours in which women formed the majority. The Kraft durch Freude (KdF, Power through Joy) organization became the largest travel organization in the world.

In 1923 350,000 French visited Algeria and Tunisia, a flow for which France built an extensive railway network (while at the same time forcing the indigenous 'obstinate vagabonds' to settle). When in 1936 the left-wing Popular Front started to shape tourism as a national project, we can observe a convergence between socialist, democratic, and fascist regimes to a common Western set of values of social prestige, familial and national unity, return to nature, and regeneration of the body. But the share of the car in international European tourism was modest: 5 percent for Britons going abroad in 1930, 10 percent of the train tourists coming and going to Italy. Only in Switzerland, from 1933, did more tourists enter the country by car than by train. No doubt car tourism formed a decisive factor in the eventual dominance of road over rail. Looking at all the car mobilities in both continents during the Interbellum, we can estimate that one quarter consisted of pure pleasure, one quarter of pure business, with the remaining half a mixture of pleasure, utility, and business use.

If we, however, wish to sketch the subtle changes in this adventurous 'mood' of pure pleasure, we need travelogues and novels, fictional and semi-fictional sources. There were plenty: a real explosion of travel accounts occurred during the Interbellum. There was the 'lost generation' of British young men, members of the male elite, many of whom had been killed during the war in disproportionate numbers. In a way they are comparable to the Belgian-French *avant garde* from the previous period because they, making "a religion out of speed and movement," became an example for their younger colleagues, as they "made us fear boredom more than death."[43] Henry Vollam Morton (who wrote an impressive array of travelogues, first on British regions, then on the Middle East) and D. H. Lawrence (who traveled with his wife through Sardinia in a motor bus) formed the tip of a gigantic iceberg of middle- and low-brow thrilling and adventure novels (such as the Biggles series by 'Captain W. E. Johns'). One of the youngsters was Graham Greene, "who had missed the enormous disillusionment of the First World War, so we went looking for adventure."[44] For Germany, Andrea Wetterauer analyzed no fewer than 711 travel accounts appearing between 1923 and 1929 in the *Frankfurter Zeitung*, and characterized them as a new "experience tool" of an insecure middle class, focusing on the "automotive world of the senses" rather than on the surrounding landscapes.[45]

Furthermore, the British explorers-by-car we saw appear in African colonies earlier in this chapter formed the tip of an iceberg of international adventurers, such as the Germans Klaus and Erika Mann or Rose Wilder and Helen Dore Boylston's journey from Paris to Albania in a Model T. There were many women travelers among them, and the literature does not agree whether "a separate genre of female travelogue cannot be distinguished" (as Karolina Fell opined in her overview of German car travelers) or whether there exists a separate "écriture feminine" (*sic*) in which gender is written in the "fissures" of the texts, one of the topics being women's "sexual integrity . . . in the encounter . . . with indigenous males."[46] For our purpose Georgine Clarsen's conclusion about women's adventures on long-range journeys is of interest, as they tended to moderate masculine violence and aggression.[47] These and many other travelogues show that the spectrum of female adventure was as variegated as that of men. Sasha Disko added an interesting dimension to these cross-continental car and motorcycle travels during the Interbellum when she analyzed these travelers' practices as Orientalist and racist, "claiming modernity that the other was purported to lack."[48]

Interbellum Western car culture was not as inclusive as this brief overview might suggest, however. The heroines in German novels about lower-middle-class women (such as Irmgard Keun's *Das kunstseidene Mädchen* [1932; The Artificial-Silk Girl] or Marieluise Fleisser's *Mehlreisende Frieda Geier* [1931; Frieda Geiger, Commercial Traveler in Flour]), members of the twelve million working women in Germany, more than anywhere in Europe, had their mobile adventure either in other people's luxury cars (Keun) or outside their 'philistine' cheap cars used for their work (Fleisser). The same is true for their male counterparts, such as Hans Fallada's 'Little Man' (in *Kleiner Mann—was nun?*,

1932). Such adventures take on a grim flavor in the case of the working class, if they are on the run from the Dust Bowl, looking for work (John Steinbeck, *The Grapes of Wrath*, 1939), or indulging in their gluttonous and fully untamed car consumption during the Depression (a cynical Erskine Caldwell in *Tobacco Road*, 1932). Bertha Thompson ('Boxcar Bertha')'s description of subversive mobility in and under railway wagons and in dilapidated Fords, as well as life in the hobo 'jungles,' shows that female hobos preferred traveling in their cars not so much out of an adventurous desire but rather out of personal safety, reminding us of the experiences while 'driving Black.' At the same time, she showed how these cars ruined the culture of (depending on the estimate) between three hundred thousand and two million tramps and transients in the United States.

For *urban motoring* we lack semi-, let alone nonfictional witnesses altogether, so here we cannot but trust that belletristic fantasizing was somehow inspired by 'real' experiences, just as we assumed this to be the case with the Expressionists of the previous period. We have, for instance, Erich Maria Remarque (pseudonym of Erich Remark)'s contributions to the popular magazine *Sport im Bild* (Images of Sports) in the second half of the 1920s, which described urban traffic, a separate world, as a war on the streets against obstacles such as pedestrians and trams. We have a very critical Communist Ilya Ehrenburg who, in his *Leben des Autos* (1929, translated much later as *The Life of the Automobile*), sees cars "laconically overrun . . . the pedestrian." Cars are "destined to annihilate the people," but as they are also a part of a worldwide system at the same time they "kill[..] Malayan coolies and Mexican workers."[49] Robert Musil also reveals in his novel *Der Mann ohne Eigenschaften* (1930–42, The Man without Qualities) that the 'car society' manifests itself first and foremost in the city, for instance through the iconic fatal accident-as-spectacular-event at the start of his novel.

But the quintessential novel on the inner life of a car commuter is Sinclair Lewis's *Babbitt* (1922), which may be seen as the automotive equivalent of H. G. Wells's cycling romances from an earlier period. Analyzed by many, for our purpose Lewis (who received the Nobel Prize for Literature in 1930) describes the car as a vehicle of conformism, the exact opposite of a contraption to deviate from the norm; the car is used to reconcile men with (feminine) consumption. In the literary canon characterized as a satire of a small-town philistine culture, but appreciated (for instance as an iconic creator of the suburb) by Lewis's contemporary middle-class compatriots, the novel depicts Babbitt's parking of his car as "a virile adventure masterfully executed." During his commute, "three and a half blocks of romantic adventure," he feels proud about his electric cigar lighter, "a priceless time-saver." The effectiveness and success of the novel's irony rested on the fact that Lewis seemed to identify partially with his protagonist, so Lewis-the-author satirized himself, and through him, his readers. "Some day I'm going to take a long motor trip," Babbitt confesses to his wife. "'Yes, we'd enjoy that,' she yawned."[50]

And yet, fictional and nonfictional travel writers during this period encountered some problems of how to combine the new ideal of family harmony with adventure. Some solved this by denial: German travel writer Colin Ross took his family on many trips, but described these trips as individual affairs. American William Carlos Williams does not mention his wife in his *A Voyage to Pagany* (1927; the trip took place in 1924), though she accompanied him during their entire journey through Europe. But the ideal-typical 'domesticated' adventurer (literally so, because the car had now become his 'home away from home') was the popular German travel writer Heinrich Hauser, who, constantly looking for a regressive form of harmony outside of the family, found its collective mobile experiences "not at all worth telling." Hauser invented a new version of 'flight': car tramping (*Autowandern*) experienced as a new slowness enabled by improved roads, a slowness that in fact was fast on the new German *autobahnen*, but experienced as passivity, "a tranquility, a recuperation [*Erholung*]." It was also a flight from the family, or perhaps from the woman as such; as a substitute, the car became a "prude lover" (*keusche Geliebte*).[51]

Several cultural historians and other students of the Depression have commented on the call for "coolness" and observed a "cool" attitude resulting from the taming of the car adventure through the cyborg mechanism (such as, for instance, to be found in Remarque's 'urban traffic' stories). Cultural sociologist Helmut Lethen observed the German intelligentsia's "cool conduct" (*kaltes Verhalten*) as a behavioral response to modernization. Max Weber (with his "scientific type of the cool persona"), Ernst Jünger (who dreamt of a new race of men with "an icy brain and a glowing heart"), Bertold Brecht (referring to the *Kältepanzer* [armor of coldness] of modern men, moving as "mobile subjects without inner depth"), Klaus Theweleit (and his analysis of the German "soldier man" who could perform a "bodily transformation" between sexuality and violence and who identified with machines, preferably the car)—they all observed a change in men's behavior in relation to their mobility.[52] And not only men: see Tamara de Lempicka's *AutoPortrait* from 1925, a painting of a 'new woman,' sitting in a Bugatti (she owned a small Renault herself), throwing a cool gaze at us as onlookers.

No less nonconformist were the German New Women writers who during the Weimar Republic discovered the car as an explicit *emancipa-*

tion tool. Cultural and media studies scholar Anke Hertling's dissertation not only gave us an excellent gendered history of German autopoetics (in which most of the elements of the automotive adventure as they appear in this book are to be found), it also presented three case studies of women publishing their feuilletons in magazines such as *Die Dame* (The Lady), *Sport im Bild* (Sports Illustrated), *Elegante Welt* (Elegance), the tabloid *Tempo*, and some Swiss periodicals, in which the car functions as one of the actors (feuilletons are a part of a semi-literary, often illustrated *Unterhaltungsliteratur* [entertainment literature]). The androgynous Ruth Landshoff-Yorck, one of the first well-known female motorcyclists in Europe, played with the affinity between fashion and car culture, driving around in an expensive car with a specially designed body, her clothes harmonized with the car's upholstery, but a coverall at the ready in case of a sudden defect. She describes automotive beauty contests in an ironic style, in which the noisy lack of skill in gear changing counts as an advantage among the male jury and the car's color "hardly influences" engine performance.[53]

By contrast, Thomas Mann's daughter Erika mimics a masculine attitude to deal with male 'help' during her high-speed trips. Like Landshoff-Yorck taking part in the Berlin *bohème* of the 1920s with its drugs and generous sexuality, she 'conquers' the whole of Europe by car with her brother Klaus, with whom she also makes an automotive tour around the world, as we saw. With Klaus she will also report on the Spanish civil war later in the decade. She follows a mechanic's course before her journey to Morocco and takes part in a rally across Europe in 1931. Hertling recounts how through Mann's first prize at the rally "the prejudices against women as car drivers are refuted," but this does not lead to an opening up of the heavily masculinized racing culture, despite the performances of several more female racers such as the famous Clärenore Stinnes (the first woman who circled the earth in a car) or Elisabeth Jinek, who invented the careful study of the racecourse before a race (and finished first in a race of all non-factory-sponsored drivers in her Bugatti).[54]

Annemarie Schwarzenbach, an admirer and close friend of both Mann and Landshoff-Yorck who socialized in and around cars as intimately as they did, does not seek adventure but uses automotive mobility as an existential expression of her suffering from the emptiness of the period. Schwarzenbach's contributions to the German and Swiss entertainment press reveal the crucial importance of the car interior (even in a cabriolet), the car-as-capsule, where gender identities can be investigated and questioned, showing at the same time the importance of literary theory's Quest of the Self as *raison d'être* of (semi-)

fictional prose. Car travels to Iraq and Persia, in the US, and (with French Olympic sail champion Ella Maillart, after acquiring contracts from press and photo agencies) to Afghanistan show how she construes her own car adventures "not like English ladies with ice-cooled bottles, accompanied by men and by boys," but in a Ford Roadster deluxe carefully prepared before her three-months-long journey into a landscape with mule tracks and sand dunes where the use of camels often proves superior.[55] What strikes the reader of her feuilletons and travelogues especially is the conscious use of photographs to picture the strangeness of the exotic for which the writer remains speechless. After the Nazis came to power in 1933, such women disappeared from the entertainment press, to make space for 'steel-hard' men like racing aces Hans Stuck and Bernd Rosemeyer.[56]

Summarizing, mainstream travelogues and novels picture a fundamental change in adventurous driving experience, not so much substituting for those of the previous period, but providing a new layer upon the car adventure, a layer of normalization (in the analysis of literary historian Jürgen Link),[57] of Foucaultian regulation of the passenger and the driver back to normalcy, with their car trip as a self-therapeutic practice. In this layer, the car adventure becomes so tamed that the vehicle becomes a kind of background noise, always there but never to blame, just like 'family' or 'marriage.' At the same time, recently unearthed car adventures produced by German New Women show, as we just saw, how for them early car driving has not in the least lost its attraction for emancipatory purposes.

But the Interbellum is also the period of a third layer of car adventurousness, already present in the previous period in utterances of *popular culture* such as film, as we saw in Chapter 1, but in this period visible in mobile 'reality' through a fledgling American 'hot rod culture' of the 1930s. But film, too: the emerging five big Hollywood studios produced race movies (from *The Roaring Road* of 1919 to *The Crowd Roars* of 1932 [with a 'cool' James Cagney]) and movies in which car speed was coupled with violence (*Scarface*, 1932). Film students who analyzed the affinity between film watching and car driving theorized and analyzed how car occupants now learned to perceive the environment in a filmic way, supported by the comfortable technology of the family capsule speeding over the improved roads. In literary utterances, and certainly not only in lowbrow versions, the "crash" became a "topos of modernity" and an accepted interruptive part of the literary plot, as literary historian Claudia Lieb showed in a book-length study of this phenomenon, thus giving a nice example of the 'affinity' of writing/reading and driving: "The accident is as much mechanical as it is linguistic."[58]

We may wonder how Western societies coped with this increasing high-speed road mobility, an increase that was as explosive as car diffusion itself. One of the ways to test that is to look at traffic casualties. Especially during the 1920s the number of traffic deaths increased precipitously, until during the 1930s they seemed to stabilize. In 1934 the number of road traffic deaths in the UK (7,300, nearly half of them pedestrians, including many children) was nearly as high as during the notorious 'peak' in 1972 (7,700; see Chapter 3). Peter Norton has shown how in the US local business interests organized massive campaigns (including civic parades and the distribution of brochures) as a response against what they feared were protests directed against automobilism as such.[59] Anecdotal evidence from European countries also suggests a brief revival of prewar anti-car sentiments, in the UK even culminating in the founding of a Pedestrians' Association.[60] But such sentiments did not lead, in Europe nor in the US, to an increase of the feeling of terror. Instead, the car underwent a process of 'normalization,' just like we have seen in the novels just cited, perhaps also aided by the realization among some well-informed non-users that slowly motorists were starting to kill each other more than pedestrians. In the US, shock therapies to scare motorists into safer behavior did not help, despite the publication of a *Reader's Digest* brochure depicting all the gruesome details of a car collision, which made it into a film in 1936. Meanwhile the presumed 'road hogs' were arrested and either accused of manslaughter or murder (in the beginning) and then sentenced by American judges to be psychologically tested for reeducation, among them a disproportional amount of Blacks and other people considered to deviate from the norm (such as homosexuals and women), as historian Daniel Albert showed in his unjustly forgotten dissertation.[61]

One of the official answers to the 'carnage' on the roads was a technical fix proposed by the transnational roadbuilding community gathered in PIARC. Although this association only started to include sessions exclusively dedicated to road safety as late as its conferences in Munich and The Hague (1934 and 1938), it straightened, widened, and paved existing roads from the very beginning of the 1920s, supported by a golden formula implemented in all motorizing countries during the Interbellum, consisting of the coupling of car tax exclusively to roadbuilding (so-called 'earmarking'), thus creating a spiraling effect of ever higher taxes and ever more sumptuous roadbuilding funding. Several motorizing countries founded transport ministries at the beginning of the 1920s, which stimulated the centralization of research and the setting up of laboratories. Most of these governments initially were against the 'anti-popular and anti-democratic' freeway building programs of fascist Italy and Germany. On a transnational level a controversy within the League of Nations also played a role: whereas International Labor Office director Albert Thomas, as an admirer of Francis Delaisi's *Les deux Europes* (1929), held a plea to connect the wealthy Western with the agricultural Eastern Europe, the League's Communications and Transit Committee followed PIARC's policy of skepticism toward these 'luxury roads' (according to the Swiss Touring Club) or 'great new race tracks' (Winston Churchill). The British touring club, supported by the international tourist association IAT (the successor of LIAT, since 1919), proposed a freeway corridor from London all the way to Istanbul, of which some small stretches were actually realized in Eastern Europe.[62] But when the freeway lobby (mostly consisting of private enterprises) in the course of the 1930s started to use the safety argument, the opinion within the roadbuilding community began to shift: freeways, one participant of the PIARC conference in The Hague in 1938 stated, through their separation of traffic flows, not only enhanced their speed (in a technocratic vision seen as 'efficiency'), but also their safety because of the decreased chance of meeting 'collision partners.' Freeways thus introduced railway thinking into roadbuilding, a nice example of inter-artifactual technology transfer.

One of the feats of the 'mobility turn' among historians of transport is their common establishment of a solid foundation of roadbuilding history, resulting in two new insights: roadbuilding was effective because central planning and financing were feeding an intermediary level of implementation (provinces, states, *Länder*), and secondary roads rather than freeways helped establish the 'car society' of the late 1930s. This infrastructural underpinning of the 'car society' by a double system of improved secondary roads and new high-speed roads strengthened automobilism in its struggle with the railway interests, so much so that planners started to speak of a 'coordination crisis.' The basis for this crisis was laid directly after World War I when thousands of drivers and surplus military trucks flooded the European markets, the start of so-call 'wild' national trucking and motor bus cultures, because they threatened with their smaller loads, their farm-to-city transport, and their door-to-door practices the bulk-oriented transport policy of the railroads, leading to national regulation and legislation to tame the adventurousness of the wild buses and trucks. Motor buses, whose production boomed during the 1920s, enabled a new form of collective 'folk tourism' in which women often formed a majority. Similar developments took place in the United States, but there, additionally, an openly subversive form of urban taxis (jitneys) also occurred, starting somewhat earlier (between 1910 and 1920) and driven by the unemployed (including many Blacks in the southern states). Intercity jitneys survived as Black bus lines into the postwar years but most of them had disappeared after the Great

Recession of 1929, but the 'white' Greyhound and the National Trailways Company remained.[63]

It seems that the railroads initially did not see the car as a threat (planners had given them the role of 'feeder' to the railway), even when many regional and local tramway systems started to collapse under their own weight of deficit and mismanagement, pushed over the edge by public criticism. But when business people started to choose the car in massive numbers and railways started to lose their lucrative first-class and subscription clients, it became clear that two types of societies were struggling for hegemony: a society of decentralized, light industry producing commodities that needed to be distributed in small amounts, against a 'railway society.' This process has now been well documented by a generation of transport historians.[64] In fact, the 'victory' of road over rail during the second half of the century was preceded by the demise of local and regional tramway lines. Although after the war the American variant of the generally Western process has been interpreted as a 'conspiracy' (see Chapter 4), for the Netherlands it has been shown, for instance, that many tramlines were by the beginning of the 1920s on the brink of collapse, well before the massive breakthrough of the car (but 'helped' by the emergence of the—often illegal—bus lines), due to economic problems, especially in the case of private ownership, and problems of bad management, further weakened by a lack of support from local governments.[65]

By the end of the Interbellum, most countries of the motorizing West had a fully functional road system working. They had become 'car societies' in which drivers and passengers could produce a multisensorial transcendent experience based on a haptic relationship with their car, fed by the angst-lust of the speedy trip. Mobility studies so far failed to interpret automotive experience as transcendental, not very much helped by literary theorists' focus on the car trip as a Quest of the Self. The modest adventure appeased and reconciled car occupants with the system, anarchism with order, adventure functioning as a test of safety rather than a rejection of certainty.

This is the result of our analysis of literature's *content*. Regarding its *symbolic* dimension: by transforming the automotive 'swarm' into a 'flow,' 'traffic' rather than 'car' became the dominant metaphor of this period. And as to '*affinity*': the co-construction of car society and car technology, of culture and artifact, established a structural homology between car driving and capitalism. The 1920s formed the pivotal decade in which the cultural, infrastructural, legislative, technical, and economic forces coincided to pave the way for car society's emergence. With the car society established in the West, mobility politics became car politics.

Notes

1. Yosano Akiko, *Travels in Manchuria and Mongolia; A Feminist Poet from Japan Encounters Prewar China; Translated by Joshua A. Fogel* (New York: Columbia University Press, 2001), 107. See for the following the much more extensive treatment of this period in Gijs Mom, *Globalizing Automobilism: Exuberance and the Emergence of Layered Mobility, 1900–1980* (New York: Berghahn, 2020), Ch. 1.
2. Stewart Lone, "Japan and the Age of Speed: Urban Life and the Automobile 1925–1930" (discussion paper No. IS/05/494, London School of Economics and Political Science, July 2005), 1–13, here 5.
3. Alisa Freedman, *Tokyo in Transit: Japanese Culture on the Rails and Road* (Stanford, CA: Stanford University Press, 2011), 173. Japan's early motorization is heavily under-researched.
4. Lone, "Japan and the Age of Speed," 11–12.
5. Koichi Shimokawa, "Japan: The Late Starter Who Outpaced All Her Rivals," in Theo Barker (ed.), *The Economic and Social Effects of the Spread of Motor Vehicles: An International Centenary Tribute* (London: MacMillan, 1987), 214–35, here 215; M. William Steele, "The Speedy Feet of the Nation: Bicycles and Everyday Mobility in Modern Japan," *Journal of Transport History* 31, no. 2 (December 2010), 182–209, here 185 (four million), 197 (women). On the theory of the logistic curve as a mathematical tool to represent the 'diffusion' of vehicles see Gijs Mom, "Frozen History: Limitations and Possibilities of Quantitative Diffusion Studies," in Ruth Oldenziel and Adri de la Bruhèze (eds.), *Manufacturing Technology, Manufacturing Consumers: The Making of Dutch Consumer Society* (Amsterdam: aksant, 2008), 73–94.
6. Manuela Boatcă, "Semiperipheries in the World-System: Reflecting Eastern European and Latin American Experiences," *Journal of World-Systems Research* 12, no. 11 (December 2006), 321–46, here 322.
7. Frank Dikötter, *Exotic Commodities: Modern Objects and Everyday Life in China* (New York: Columbia University Press, 2006); Thorben Pelzer, *Engineering Trouble: US-Chinese Experiences of Professional Discontent, 1905–1945* (Leiden: Brill, 2023), 146, 148, 159, 166 (cars), 170, 173, 200.
8. Yajun Mo, *Touring China: A History of Travel Culture, 1912–1949* (Ithaca, NY: Cornell University Press, 2021), 2.
9. Lao She, *Rickshaw* (Honolulu: University of Hawaii Press, 1979).
10. Vinayak Chaturvedi, "Introduction," in Vinayak Chaturvedi (ed.), *Mapping Subaltern Studies and the Postcolonial* (London: Verso, 2000), vii–xix, here vii–viii, x–xii; Gyan Prakash, "Writing Post-Orientalist Histories of the Third World: Perspectives from Indian Historiography," *Comparative Studies in Society and History* 32, no. 2 (April 1990), 383–408, here 400 ("counterinsurgency"), 400n34 ("against the grain").
11. David Arnold, "Subaltern Streets; India, 1870–1947," in Tariq Jazeel and Stephen Legg (eds.), *Subaltern Geographies* (Athens: University of Georgia Press, 2019), 36–57, here 42.
12. Thomas Williamson, "The Fluid State: Malaysia's National Expressway," *Space & Culture* 6 no. 2 (May 2003), 110–31.
13. Patrick Manning, *Slavery, Colonialism and Economic Growth in Dahomey, 1640–1960* (Cambridge: Cambridge University Press, 1982), 182.
14. Gijs Mom, *Atlantic Automobilism: Emergence and Persistence of the Car, 1895–1940* (New York: Berghahn, 2015), 436–37.

15. See the contributions in Jan-Bart Gewald, Sabine Luning, and Klaas van Walraven (eds.), *The Speed of Change: Motor Vehicles and People in Africa, 1890–2000* (Leiden: Brill, 2009).
16. Salvatore Attardo, "Irony as Relevant Inappropriateness," *Journal of Pragmatics* 32, no. 6 (2000), 793–826, here 794. The following paragraph is taken, with some adjustments, from Mom, *Globalizing Automobilism*, 10–11.
17. Rebecca Clift, "Irony in Conversation," *Language in Society* 28, no. 4 (1999), 523–53, here 535; Christian Burgers and Margot van Mulken, "Het ironisch spectrum: Een overzicht van onderzoek naar het begrip en de retorische effecten van verbale ironie," *Tijdschrift voor Taalbeheersing* 35, no. 2 (2013), 183–202, here 184 ("wolves," "sheep").
18. Andrew Denning, *Automotive Empire: How Cars and Roads Fueled European Colonialism in Africa* (Ithaca, NY: Cornell University Press, 2024), 17, 274–78, 281 (Huillery).
19. Joshua Grace, *African Motors: Technology, Gender, and the History of Development* (Durham, NC: Duke University Press, 2021), 9.
20. Ibid., 40, 49, 60.
21. See, also for the following, Mom, *Globalizing Automobilism*, 154–66.
22. John Brian Freeman, "Transnational Mechanics: Automobility in Mexico, 1895–1950," unpublished dissertation (New York: City University of New York, 2012), 60.
23. Ricardo D. Salvatore, "Early American Visions of a Hemispheric Market in South America," in Berndt Ostendorf (ed.), *Transnational America: The Fading of Borders in the Western Hemisphere* (Heidelberg: Universitätsverlag C. Winter, 2002), 45–64; Fiona Wilson, "The Conflict between Indigenous and Immigrant Commercial Systems in the Peruvian Central Sierra, 1900–1940," in Rory Miller (ed.), *Region and Class in Modern Peruvian History* (Liverpool: University of Liverpool, 1987), 125–61.
24. Shawn William Miller, *The Street is Ours: Community, the Car, and the Nature of Public Space in Rio de Janeiro* (Cambridge: Cambridge University Press, 2018), 4.
25. Ibid., 10, 15, 34, 108, 111, 177, 187; Jean Comaroff and John L. Comaroff, *Theory from the South or, How Euro-America is Evolving toward Africa* (London: Routledge, 2012).
26. See for the following Mom, *Atlantic Automobilism*, Ch. 5, 4, 6, and 7.
27. Gijs Mom, *The Electric Vehicle: Technology and Expectations in the Automobile Age* (Baltimore: Johns Hopkins University Press, 2004), 264–70.
28. For more details see Gijs Mom, *The Evolution of Automotive Technology: A Handbook* (Warrendale, PA: SAE International, 2014). On sound as an element in car design see Karin Bijsterveld et al., *Sound and Safe: A History of Listening Behind the Wheel* (Oxford: Oxford University Press, 2014).
29. Joshua Richard Butt, "Aspects of the Automobile's Diffusion in the North-West of England 1896–1939," unpublished dissertation (Manchester: Manchester Metropolitan University, 2018), 85, 87–88 (O'Connell quoted on 19n38).
30. Ibid., 83.
31. Rüdiger Rabenstein, "The History of German Workers' Cycling Association, Solidarity," *Cycling History 11* (Proceedings, 11th International Cycling History Conference, 23–25 August 2000), 160–67, here 164.
32. Peter Bailey, "White Collars, Gray Lives? The Lower Middle Class Revisited," *Journal of British Studies* 38 (July 1999), 273–90, here 278–79, 285 (the bohemianism

quote stems from Stanley Pierson, *Marxism and the Origins of British Socialism* [1973]).

33. Christopher Thomas Potter, "An Exploration of Social and Cultural Aspects of Motorcycling During the Interwar Period," unpublished PhD dissertation (Newcastle: University of Northumbria, August 2007), 18, 32, 43, 190 ("tinker"), 194–95, 204, 209 ("independent"), 229, 233 ("dominated").
34. Gijs Mom and Ruud Filarski, *Van transport naar mobiliteit: De mobiliteitsexplosie (1895–2005)* (Zutphen: Walburg Pers, 2008), 79.
35. Lewis H. Siegelbaum, "Soviet Car Rallies of the 1920s and 1930s and the Road to Socialism," *Slavic Review* 64, no. 2 (Summer 2005), 247–73; Lewis H. Siegelbaum, "Roadlessness and the 'Path to Communism': Building Roads and Highways in Stalinist Russia," *Journal of Transport History* 29, no. 2 (2008): 277–94. For mixed roads see Mom, *Atlantic Automobilism*, 584; "roundups": Lewis H. Siegelbaum, "The Impact of Motorization on Soviet Society after 1945," in Manfred Krieger, Ulrike Gutzmann, and Dirk Schlinkert (eds.), *Towards Mobility: Varieties of Automobilism in East and West* (Wolfsburg: Volkswagen AG, 2009), 21–29, here 22; elite: Frank Parkin, "Class Stratification in Socialist Societies," *British Journal of Sociology* 20, no. 4 (December 1969), 355–74, here 356; Stalin quotes in Lyubomir Pozharliev, *The Road to Socialism: Transport Infrastructure in Socialist Bulgaria and Yugoslavia (1945–1989)* (Göttingen: V&R Unipress, 2023), 77.
36. Elvira Khairullina and Luis Santos y Ganges, "Tram, Trolleybus and Bus Services in Eastern-European Socialist Urban Planning: Case Studies of Magdeburg, Ostrava and Oryol (1950s and 1960s)," *Journal of Transport History* 42, no. 1 (2021): 26–57, here 30.
37. Gijs Mom and Diekus van der Wey, "Ethanolverdamping en de dieselmotor: Literatuuronderzoek naar en ontwerp van een ethanolverdamper t.b.v. het gecombineerd gebruik van dieselbrandstof en ethanolgas in een DAF-dieselmotor" [Ethanol evaporation and the diesel engine: Literature research on and design of an ethanol evaporator to combine diesel fuel and ethanol gas in a DAF diesel engine], Bachelor of Science thesis (Apeldoorn, The Netherlands: HTS-Autotechniek, 7 June 1982).
38. Virginia Scharff, *Taking the Wheel: Women and the Coming of the Motor Age* (New York: UNM Press, 1991), 66.
39. Mia Bay, *Traveling Black: A Story of Race and Resistance* (Cambridge, MA: Belknap Press, 2021), 118, 109 ("racialized"), 115–16, 121, 125, 129, 135, 149; Gretchen Sorin, *Driving While Black: African American Travel and the Road to Civil Rights* (New York: Liveright, 2020), 32, 66.
40. Sean O'Connell, *The Car and British Society: Class, Gender and Motoring, 1896–1939* (Manchester: Manchester University Press, 1998), 35.
41. Colin G. Pooley and Jean Turnbull, "Commuting, Transport and Urban Form: Manchester and Glasgow in the Mid-Twentieth Century," *Urban History* 27, no. 3 (2000): 360–83, here 363.
42. Warren James Belasco, *Americans on the Road: From Autocamp to Motel, 1910–1945* (Cambridge, MA: MIT Press, 1979).
43. Malcolm Cowley, *Exile's Return: A Literary Odyssey of the 1920s* (New York: Penguin, 1994 [1934]), 38–39; Robert Wohl, *The Generation of 1914* (Cambridge, MA: Harvard University Press, 1979), 227.
44. Quoted in Paul Fussell, *Abroad: British Literary Traveling between the Wars* (Oxford: Oxford University Press, 1980), 70.

45. Andrea Wetterauer, *Lust an der Distanz: Die Kunst der Autoreise in der "Frankfurter Zeitung"* (Tübingen: Tübinger Verein für Volkskunde, 2007), 136.
46. Karolina Dorothea Fell, *Kalkuliertes Abenteuer: Reiseberichte deutschsprachiger Frauen (1920–1945)* (Stuttgart: J. B. Metzler, 1998), 247; Mary Suzanne Schriber, *Writing Home: American Women Abroad 1830–1920* (Charlottesville: University Press of Virginia, 1997), 11.
47. Georgine Clarsen, "Machines as the Measure of Women: Colonial Irony in a Cape to Cairo Automobile Journey, 1930," *Journal of Transport History* 29, no. 1 (March 2008): 44–63, here 48.
48. Sasha Disko, "'The World is My Domain': Technology, Gender and Orientalism in German Interwar Motorized Adventure Literature," *Transfers* 1, no. 3 (Fall 2011): 44–63, here 47.
49. Ilja [Ilya] Ehrenburg, *Das Leben der Autos* (Amsterdam: Paco-Press, 1973 [1929]), 235–36.
50. Sinclair Lewis, *Babbitt* (Harmondsworth: Penguin, 1985 [1923]), 46–47, 77, 110–11.
51. Heinrich Hauser, *Fahrten und Abenteuer im Wohnwagen* (Stuttgart: DoldeMedien, 2004 [1935]), 31; Heinrich Hauser, *Schwarzes Revier* (Bonn: Weidle, 2010 [1930]), 12–13; Gregor Streim, "Flucht nach vorn zurück: Heinrich Hauser—Portrait eines Schriftstellers zwischen Neuer Sachlichkeit und 'reaktionärem Modernismus,'" *Jahrbuch der Deutschen Schillergesellschaft* 43 (1999): 377–402, here 380.
52. Helmut Lethen, *Cool Conduct: The Culture of Distance in Weimar Germany* (Berkeley: University of California Press, 2002), 12–13; Klaus Theweleit, *Männerphantasien 1 + 2* (Munich: Piper, 2009 [1977 and 1978]), 487.
53. Quoted in Anke Hertling, *Eroberung der Männerdomäne Automobil: Die Selbstfahrerinnen Ruth Landshoff-Yorck, Erika Mann und Annemarie Schwarzenbach* (Bielefeld: Aisthesis, 2013), 151.
54. Ibid., 187.
55. Ibid., 252–53.
56. Ibid., 265.
57. Jürgen Link, "Maintaining Normality: On the Strategic Function of the Media in Wars of Extermination," *Cultural Critique* 19 (Fall 1991): 55–65, here 62.
58. Claudia Lieb, *Crash: Der Unfall der Moderne* (Bielefeld: Aisthesis, 2009), 315.
59. Peter David Norton, *Fighting Traffic: The Dawn of the Motor Age in the American City* (Cambridge, MA: MIT Press, 2008), 147–48.
60. Keith Layborn with David Taylor, *The Battle for the Roads of Britain: Police, Motorists and the Law, c.1890s to 1970s* (New York: Palgrave Macmillan, 2015), 4.
61. Daniel M. Albert, "Order out of Chaos: Automobile Safety, Technology and Society 1925 to 1965," unpublished dissertation (Ann Arbor: University of Michigan, 1997).
62. Alexander Badenoch, "Touring between War and Peace: Imaging the 'Transcontinental Motorway,' 1930–1950," *Journal of Transport History* 28, no. 2 (September 2007): 192–210.
63. Bay, *Traveling Black*, 153, 162.
64. For an overview see Gijs Mom, "Clashes of Cultures: Road vs. Rail in the North-Atlantic World during the Inter-War Coordination Crisis," in Christopher Kopper and Massimo Moraglio (eds.), *The Organization of Transport: A History of Users, Industry, and Public Policy* (London: Routledge, 2015), 8–31.
65. Mom and Filarski, *Van transport naar mobiliteit*, 333–59.

CHAPTER 3

Decolonization, Development, and the Globalization of the Car Society, 1940s–1970s

When the young James Ballard, immediately after the 'Pacific War,' entered the UK from Shanghai where he was born, one of the first surprises he experienced in the new world were the small cars ("black perambulators") he had never seen before.[1] As the post-World War II counterpart of the Japanese transcultural traveler Natsume Sōseki (see Chapter 1), and soon to become the iconic masthead of the middlebrow science fiction genre's New Wave, specializing in linking cars, violence, and sex, Ballard had admired the Packards and Buicks of the victorious Americans.[2]

In this chapter, we see the car worldwide enter into a transitional phase, in which the postwar automotive exuberance of the West is accompanied by the first aggressive efforts to export the 'car society' over the globe, not so much by exporting mobile artifacts (that came later, contrary to what we know about the spread of the car society in the West), but through 'development-by-roadbuilding.' With the roads a desire for 'modernization-through-mobility' came along. The question then is which elements of the increasingly fragmented car culture as we saw emerge in the West fitted into the fledgling car societies in the Global South and other 'peripheries' such as the Global East: a mainstream, somewhat 'flattened' disposition; a no less mainstream but still adventurous emancipatory subculture developed by Blacks, queers, women, and their passengering children; or the over-the-top adventurous and carnivalesque flight forward raging in the popular culture of film and songs, but also among hot-rodders and lowriders? We will start looking for an answer in Asia and the Soviet Union, return then to the West to witness (like contemporary inhabitants of the Global South and East) 'car exuberance' and its fragmentation emerge (using literature, songs, and film to get a better grip on its user culture), see the World Bank's development-through-roadbuilding reach Latin America and Africa, emulated by Japan in Asia, and finally visit beginning road motorizations in some countries in the Global South.

World War II, which for China lasted eight years and took fourteen million lives, ended with the surrender of Japan, not only because of the dropping of the American atom bombs, but also because of the invasion of the Soviet Union in Japan's economically crucial colony Manchukuo (Manchuria).[3] A civil war followed, in which the Communists destroyed ten thousand miles of railway track and captured two thousand excellent American and Japanese trucks from the Guomindang, before they founded the People's Republic in 1949, while the latter fled to Taiwan. Backed by the Soviet Union (and with the United States as the new enemy because of the Korean War of 1950–53), under its First Five-Year Plan (1953–58) the new government invested heavily in transport infrastructures, while at the same time strengthening its ties with the thirty million Chinese in the diaspora through its membership of the Bandung Conference of nonaligned countries in 1955. The revolution triggered enormous migration waves to the cities, despite the *hukou* regulation (a household registration system meant to avoid migration to the cities, introduced in 1958), and the famine of the Great Leap Forward (1958–61, related to the Second Five-Year Plan) cost at least twenty million lives. It is nonetheless remarkable that these same years saw a steep increase in the building of roads, considered, Soviet-style, as a supplement to the railroads, a phenomenon hardly discussed in Anglophone sources. By the end of the Maoist period (Mao Zedong died in 1976) investment in transport was still very high (one-third of GDP, double the share of 1952). Before the withdrawal of the Soviet advisers in 1960 a car and truck industry in northeastern Changchun had been set up, producing Jiefang (Liberation) trucks, Hongqi (Red Flag) limousines, and Dongfeng (Eastern Wind) middle-class cars. In Shanghai a sedan (Phenghuang, Fenix) appeared, based on a Mercedes Benz model, but until 1983 only sixty privately owned passenger cars were registered in China. By contrast, bicycle production amounted to four million by the beginning of the 1970s, its factories located in nearly every province. What characterized Chinese mobility in the 1950s was its extreme layeredness, with traditional transport contributing half of the country's GDP and around three-quarters of all transport taking place by boat. Modernization was layered too: it also took place through the production of ball bearings and pneumatic tires for wheelbarrows and hand-pulled and animal-draft carts.

This is not to say, however, that adventurousness was absent in China's early postwar mobility history. During the 'ten year disaster' of the Cultural Revolution (1966–76) a form of free 'revolutionary tourism' by train emerged, whereas in the cities Red Guards confiscated jeeps and trucks to drive around "like mad." These youngsters' memoirs are replete with adventures, including stories about forms of "propaganda truck warfare" and gigantic flows of girls on bicycles when a factory workday in Shanghai ended.[4] Some of them redid the Long March on foot. 'Barefoot doctors' were often the only village inhabitants owning a bicycle. The revolution raged mostly in the cities (although thousands of 'sent-down youths' ended up in a harsh country life), where especially intellectuals and party officials suffered from the movement's anarchy and cruelty. Remarkably, during this mobility carnival the 400,000 'vehicles' on the road in 1970 alone caused no fewer than 55,400 'road crashes.'

There are many similarities with India's motorization, or lack thereof. Like China, India started its independence (1947) with famines and a civil war between Muslims and Hindus, triggering enormous migration and refugee waves either on foot or on bullock carts, but also in (or on the roof of) trains. And like China, India started ambitious intercity roadbuilding as conceived in the Nagpur Plan of 1943, although Partition (the violent creation of Pakistan and Bangladesh) reduced the length of the 'Indian' network by one-third. Until 1961 India was the largest user of World Bank funds, received in the context of the Cold War against the Soviet Union, whose model of industrialization India started to follow from the mid-1960s, including the nationalization of trucking and busing (the railways were nationalized in the colonial period, in 1923). Ford and General Motors left the country and national manufacturers took over, including the automobile division of Tata (cooperating with Daimler-Benz). West Bengal's Hindustan Motors' Ambassador (based on the British Morris Oxford, and used as an example of 'creole technologies' by David Edgerton[5]) became the symbol of India's modernity (for the middle classes, because Bajaj Tempo started production of light commercial vehicles in 1958 under license, among others, from a German manufacturer, which later would evolve into modernity's symbol for the masses, in the form of the auto-rickshaw).[6] But the 9 million bullocks and the 1.5 million bicycles (increased to 30 million in 1974) formed the backbone of India's layered freight and passenger mobility, so much so that—a unique phenomenon in the postwar motorizing world—India's share of the car in the modal configuration of motor vehicles *decreased*, to give room to a motorcycle fleet, which increased until far into the next century. The tension between the two rival Soviet and Western models prompted India to develop its own roadbuilding policy for which it had to seek support elsewhere. It found this in the UN ECAFE (Economic and Social Commission for Asia and the Far East, of the United Nations, later to be renamed ESCAP, the 'S' standing for Social and the 'P' for Pacific) as well as in the PIARC, of which it became a member together with other former colonies Indonesia, French Indochina, Thailand, and Cambodia. In both organizations India actively contributed studies on 'low-cost roads.'

The Soviet Union was not only an example to follow for nonaligned countries. Even in the West some transport economists looked in awe to the country's relation between traffic growth and output growth (which was higher than that of the US), a feat nearly fully realized through its extensive railway network.[7] Its notorious 'roadlessness' (see Chapter 2) and its policy of a "reasonable limitation" of cars only began to be corrected during the 1970s, when the agricultural sector was developed, and a system of trucking and busing emerged. Different from its Eastern European neighbors, the SU did not know many urban tramway networks, so when cities started to expand Soviet urban planners often opted for trolley networks, or a combination of trolleys and diesel buses. In Eastern European cities existing tramway systems were in the same transition period (1966–75) often expanded with bus lines, except in Czechoslovakia, which followed the Soviet example.[8] By 1970 the Soviet Union counted 2,000 bus systems and when railway-poor Central Asia started to become developed, trucking over new roads in the Soviet states Tajikistan, Kyrgyzstan, Turkmenistan, and Kazakhstan emerged. Still, as late as 1990 half of all freight (in ton-km) and 40 percent of all passengers (in passenger-km) were hauled by the Russian railways.

A focus on 'scarcity' (against a background of an emphasis on modernization theories, and later consumption history), based on the analysis of the 1980s by the Hungarian economist János Kortai in his *Economics of Shortage*, has led mobility scholars of Eastern Europe and the Soviet Union to depict this world region's car culture as a near-Western variety (approaching socialist societies primarily as 'lacking' something), treating Nikita Khrushchev's proposal of a collective form of car use during his secret anti-Stalin speech at the 20th Party Congress in 1956 (and a later speech in Vladivostok in 1959) as a mere footnote.[9] Well before his successor Leonid Brezhnev's decision to follow the West's mass motorization in the mid-1960s (followed by leaders of Eastern European countries), and shortly after his visit to the US, Khrushchev's idea of "public taxi pools" (*prokat*) seems to have resonated among the Soviet population, according to Luminita Gatejel (who dedicated an entire study to the crucial Soviet debate about individualist versus collectivist mobility). At that moment the USSR counted one million passenger cars (mostly owned by state organizations). And although hundreds of car rental outlets were founded in the early 1960s, the consensus among Soviet historians is that the plan failed because the fleets consisted of secondhand cars in bad technical condition. Still, the national newspaper *Literaturnaia Gazeta* as late as 1971 contained articles questioning "whether the personal automobile did fit into socialist society."[10]

There was a similar effort in Hungary, undertaken by the top of the Hungarian Socialist Workers' Party: although, according to György Péteri, the idea of "a gigantic system of car sharing that encompassed the whole nation" was never discussed in the Hungarian public realm, a large contingent of the state's 873 passenger cars (1971) were "integrated into a collective system of central coordination and car sharing." The high party cadres, however, undertook an illegal "sneaking privatization" of such cars amounting to nearly one-fifth of the total. When a government resolution in 1972 restricted the use of the cars for private use, members started to buy their own cars whereupon the party decided to reimburse them for official use. When the party then started to encourage the official use of privately owned cars, a car elite was created, whereby others had to wait for months or years "after paying in full for the car." In another publication, Péteri (who considered Khrushchev's proposal "feasible") showed how the cartoons in a Hungarian satirical weekly criticized the car fetishism of this elite.[11]

The idea of a "democratic turn" (in that every citizen could borrow a car at an affordable price) also resonated among certain leading circles in other Eastern European countries, a debate that lingered on way beyond the succession of Khrushchev by Brezhnev in 1964 (this is a phenomenon we need much more research about, now that alternatives to the current car society are being discussed more intensely [see Chapter 5]). But in the German Democratic Republic another trend interfered, inspired by West Germany's car exuberance and building upon the Horch and BMW factories within the GDR's territory with the alternative design of the Trabant and Wartburg and their plastic bodies, their two-stroke engines, and their 'basic' interior equipment. The GDR set an example of a privately owned 'socialist car,' which may have inspired Aleksey Kosygin (chair of the Soviet planning commission with a background in the light industry) to launch his plan for a 'Soviet car,' a Fiat-clone Lada produced at a new plant in Togliatti (1970). Kosygin's Romanian colleague presented a similar plan. Both plans were based on the acquisition of Western technology, similar to the 'borrowing' of urban planning knowledge in the same period. Kosygin proposed in the eighth Five-Year Plan (1966–70) to quadruple national car production. But while public transport increased two and a half times faster than urban population,[12] as far as we know there did not emerge a somewhat massive car culture beyond privileged party members and government officials before the first years of the 1970s, when a "family car" (the Lada-Zhiguli) rolled out from the factory line at Togliatti.[13] By the mid-1970s, national production had risen to 1.2 million, disproportionally owned by the 'intelligentsia.' The result in this period was a "curious mix of a Stalinist system of privileges, collective use and private ownership."[14] Gatejel attempted to define a "genuine socialist car culture," characterized by the production of "one model for each class" (instead of "one car for everybody"), produced in a system with "special retail and distribution patterns" with "privileges enjoyed by the elite" and a "special role [for] mechanics."[15]

It is indeed remarkable that Soviet planners in this period never envisaged a car *system* (with a road and car maintenance infrastructure). Even when in 1971 the Lada factory in Togliatti started mass production of passenger cars, it did not set up a dense nationwide repair and maintenance network, enabling Soviet mobility historians to spend their energy on descriptions of an informal maintenance system based on peri-urban garages meant for masculine homosocial bonding and car repair. But Soviet scholars *did* come up with a second explanation (next to 'scarcity') for the rental system's failure: the conscious creation of an expectation, by the ruling elite, of an impending 'car society,' which had been envisioned already by Stalin in the 1930s (see Chapter 2) and was revived under Brezhnev. There was a *car desire*, testified by the more than two million copies of each monthly issue of *Za rulem* (Behind the Wheel). But although in the SU the production of commodities between 1955 and 1965 increased tenfold (from TVs and refrigerators to cars), the annual number of cars in private ownership remained the same, and very low, by far most of them to be found in Moscow, with its 70,000 private car owners in 1963. In 1973 their number had increased to 250,000, many of them used in an informal economy as illegal taxis. By 1970 only 2 percent of Soviet households owned a car, but such a national average might easily make us miss the visibility of car traffic in the large cities.[16]

A much richer variety of scholarship enables us to sketch quite a diverse picture of the 'layering' of mobilities in Asian societies. An excellent overview of Southeast Asian transport by geographers Howard Dick and Peter Rimmer points at the 'fundamental dualism' in these societies, resulting in the emergence of 'two mobilities,' dominated by a "slow diffusion of technology to the hinterlands," resulting in "the rapid diffusion of electric tramways" in the cities versus the "slow diffusion of the railways."[17] Layering, however, was not expressed in two, but in *multiple mobilities*, which becomes perfectly clear in by far the best overview, also written by Peter Rimmer, of the emergence of a new layer of semi-public transport carried by mopeds, motorcycles, especially designed mini versions of trucks and buses, three-wheelers, and mopeds with sidecars, a layer that forms the intermediary between the old city and the new, the slums and the boulevards.[18] In a brilliant, more recent analysis of India's "bazaar like urbanism" Tim Edensor and Mark Jayne added the perspective of distinctive non-Western, haptic, and intense olfactory and acoustic characteristics of urban mobilities, enabling traffic participants to "weav[e] [their] path" through what Westerners and local elites experienced (and often still experience) as 'chaos.'[19]

Japan, which lost two million citizens during the war, showed a quite different motorization pattern. During the seven-year reign by the American army, Japan's unprecedented growth, stimulated by its function as a supply base during the Korean War, catapulted the country into the third-largest economy in the world (its growth during the 1960s "exceptional by world standards"),[20] not least through the export of its passenger cars, the production of which increased from 162,000 in 1955 to eleven million in 1980. A road network after an American pattern (a breach with its freeway planning of the 1930s when Nazi Germany figured as design example) was built before demand. But the true "unsung hero of Japan's postwar recovery" was the bicycle and even the rickshaw, including heavier versions used for freight transport and taxi functions. As late as 1952 bicycles still formed 87 percent of the registered vehicles; cars took only 6 percent, after horse-, ox-, and handcarts with 7 percent. By 1970 thirty million bicycles were counted, used by nearly all men and most of the women. A typical Japanese version of the bicycle at the end of the 1950s made Japan "into a nation of bicycle citizens": the *mama-chari* (mother's chariot) was an easy to ride and cheap city bike that had captured half of all sales by the mid-1960s, when a second type, the mini-bike, was introduced, representing 40 percent of sales in 1973.[21]

In 1980 the passenger car's performance (in passenger-km) overtook the train. That was the result of the decision to build Japan's own version of the 'car society,' heavily benefitting from the orders by the American army during the Vietnam War. Ezra Vogel's famous study of the emergence of the *sarariiman* (salaryman) and his family showed how a 'middle-class consciousness' became directly related to the possession of a home (with a TV, a refrigerator, and an air conditioner) and a small car.[22] The *maika gannen* (my-car era) started in 1966, when there were ten million of them, despite the restrictive government policy of limiting household ownership to one car maximum, as well as "parking requirements, fuel and automobile taxes, driver licensing requirements, and the auto inspection program."[23]

When Japan started to support modernization-through-mobility in other Asian cities (dealt with in more detail later in this chapter), the struggle between Western and local interests resulted in a rich variety of layered solutions. To give only one example of this transitional period toward a full-blown coordination struggle between rail and road and between different road vehicle proponents, in Philippine Manila it was the struggle between the government, backed by the World Bank, and the illegal services of minibuses and motorized tricycles that resulted in the proliferation of the former (called *jeepneys*, a name derived from the wartime American jeep and the prewar jitneys) and the demise of the cycle rickshaws. Adorned with a homelike feel through the addition of curtains and religious paraphernalia, the jeepneys' public image was determined by decorative paintings, which made the vehicles into 'customized folk art on wheels.' Under the influence of a middle-class critique of being *bakya* (lower-class, rural taste), jeepneys were slowly replaced by air-conditioned vans from Korea and Japan.

Up to now, we have dealt with global mass motorization as a concatenation of national histories. In order to understand the worldwide emergence of a car society in this period, however, we have to analyze the culture of car exuberance emerging in the United States after World War II, expressed in the production in the 1950s of 'a billion miles a day on tyres,' and establish how it influenced the globalization of the 'car society.'[24]

America's "great leap forward," as Giovanni Arrighi called it in his brilliant economic analysis of the 'long twentieth century,' can be seen as a kind of global New Deal, in which "aid to . . . poor nations would have the same effect as social welfare programs within the United States—it would give them the security to overcome chaos and prevent them from turning into violent revolutionaries." This was implemented through the World Bank and IMF, which functioned as "the US Federal Reserve System" (with the IMF as the "Ministry of World Finance"), led not by banks but by the American government characterized by their "orientation towards Wall Street and an investment-banker approach to development," their top personnel coming "directly from the New York financial community." It was also done through many large vertically integrated corporations, not the least of which were producing cars. They "expanded transnationally as soon as they had completed their domestic continental integration [becoming] so many 'Trojan Horses' in the domestic markets of other states." A third factor was the Marshall aid to rebuild Europe, and when that did not seem to be enough to create a world market for American goods, a Cold War, based on fear of Communism, enabling a continuous rearmament race, and culminating in the Korean War in 1953.[25]

Domestically, this "golden Age of Capitalism" started with Operation Abundance, a $100 billion program, based on the GI Bill of 1944, targeting the nine million war veterans, who became 'the most privileged generation in American history' with special housing mortgage benefits. This jump-start into a welfare state, fed by an economic growth related to the Korean War and the rearmament race with the Soviet Union, created two 'suburban societies,' the blue-collar version derogatorily called 'trailer trash' (by the early 1950s the US counted 12,000 trailer parks). Both types of low-density suburbs with their predominately white nuclear families formed the very basis of American car exuberance. Typically, working mothers who brought their children to school by car (later to be called 'soccer moms') developed a new style of car culture by linking such trips with shopping during their commute.

However, as late as 1968 still half of all worker families surveyed in 1957 did not own a car, a share that only diminished to 32 percent a decade later. 70 percent of white families still commuted by public transport. President Lyndon B. Johnson's $15 billion War on Poverty during the 1960s not only enhanced the poor's participation in the consumer society through the support of mass transit, it also introduced a nationwide 'busing' program to bring their children to and from school.

But the most remarkable trend in the Western car use profile was the growing share of long trips, mostly for pleasurable purposes. And although during the 1970s aviation further increased the distance to vacation destinations, the car remained dominant, at least in the number of trips. In the US such trips were realized in cars consciously designed with 'comfortable,' very soft suspensions and, seen from a European perspective, giant proportions (in their chrome-clad body shape looking like oversized jukeboxes), impractical for everyday use (especially in urban traffic); their worldwide sales were consciously jeopardized in favor of a domestic celebration of affluence, hedonism, and a carnivalesque hysteria of greed. The longer trips were produced on an Interstate Highway System of freeways built from the end of the 1950s to the mid-1970s, 66,000 km long and not only allowing affluent nuclear families to tour in massive numbers to their holiday destinations (especially sunny California, where the 'dude camps,' with their nostalgic celebration of the Western, were popular), but also to 'clear the slums' because a large proportion of them crisscrossed the American inner cities, a racialized policy because many of the affected neighborhoods were inhabited by Black people. An early critic of this system was Lewis Mumford, who saw its highways "slashing through old neighborhoods, stealing land from public parks, dumping traffic in urban centers," thus prefiguring later roadbuilding resistance.[26] Although the motive for this megalomaniac construction was economic growth, it was 'sold' to Congress "in terms of national defense and safety," first by car lover President Truman, then by his successor President Eisenhower.[27]

This phase has been analyzed along two different lines of argumentation, represented here by Avner Offer's study of the "challenge of affluence" and Shelley Nickles's analysis of car marketeers' motivation research. Whereas Offer emphasized the "sensual gratification" rather than social distinction as the very characteristic of this classless myth of the middle-class society, Nickles intended to bring class back into consumption history by showing how working-class consumers "moved up economically but not culturally."[28] The preference for large powerful cars with tailfins and chrome strips was not an expression of bad taste or status-seeking or the result of manipulation, but a "distinctive working-class culture created not by men but by women."[29] Nickles's analysis 'defuses' those middle-class observers that did not manage to think beyond their own class base, one of them being Theodor Adorno, who compared the Cadillac unfavorably with the "superior" Rolls-Royce.[30] Instead, the middle-class culture got 'proletarianized,' just like the youths' counterculture of the 1950s and 1960s became mainstream, as we will see later.

But middle-class car culture, as we have seen it emerge in the West in the previous two chapters, not only absorbed (and in the process transformed) the subculture of countercultural rebels, other groups also joined the mainstream car drivers. American women, for instance, formed 40 percent of all licensed drivers by 1964 and they bought 25–30 percent of all cars, a phenomenon apparently so spectacular that it made transport historian Margaret Walsh somewhat prematurely conclude that the car had stopped being masculine; the industry only started to address women consumers in earnest (apart from gimmicks such as Chrysler's Dodge La Femme with "pink trim[,] a matching rain hat, umbrella, and handbag") after the energy crisis of 1973, when the masculine car market threatened to crumble.[31]

Blacks, too, joined the car exuberance. Although the Interstate system destroyed entire Black neighborhoods, it also formed an extra protective capsule around the Black motoring middle class. According to American studies scholar Cotten Seiler already shortly before the war they used *The Negro Motorist Green Book* (initiated by Victor Green) to "escape Jim Crow."[32] Like unsafe cars (see the next chapter) segregation was seen in a Cold War setting as an argument that "was driving African nations towards the Communist Bloc."[33]

Queer (LGBTQ) people used their cars for 'cruising' (and escaping from the police), whereas children, often forgotten in this expanding amalgam of automotive subjectivities, together with still many women, enriched the car culture by a new massive phenomenon: passengering, a phenomenon analyzed for the first time thoroughly by the German psychologist Rainer Schönhammer.[34]

The proliferation of autopoetic novels during this transitional period (and of autopoetic scenes in novels less concerned with mobility) testify to this *fragmentation* of the car culture and its still vivid adventurousness. We will present here only some outspoken examples of a much larger set.[35]

I agree with literary historians Deborah Clarke, Katie Mills, and Ronald Primeau that *women* (in the words of the latter) "didn't create their own unique version of America's road mythology" until the 1970s, if this is calculated from after World War II.[36] Female automotive subjectivity was represented by 'bad women,' from Ingrid Bergman's drunken driving in the movie *Notorious* (1946) to 'bad girl' Beverly Donofrio's alter ego in the novel-as-memoir *Riding in Cars with Boys* (1990, made into a movie in 2001), who clearly uses the car as a tool of personal liberation, experienced as an adventure produced in an emancipatory mood. Thrilling adventures are also produced by Anne Rolphe's protagonist in *Long Division* (1972), who drives her station wagon to Mexico to get her divorce papers, her ten-year-old daughter on the passenger seat. Freeway driving, however, is boring, and their self-centered automotive narcissism is illustrated by the lack of much landscape description. Another child passenger's adventure appears in Katherine Dunn's *Truck* (1971), fully written from the perspective of a fifteen-year-old girl, whose antidomestic and anti-nuclear family adventures end at a police station.

Whether *Blacks'* emancipatory car experiences in this period can be called adventurous is open to debate. Ralph Ellison's *Invisible Man* (1952) is already governed by a feeling of doom because of the Black chauffeur's "proximity to whiteness" (in the words of literary historian Roger Casey). John A. Williams's trip through America in a brand-new station wagon (funded by the journal *Holiday*, and published in 1966 as *This Is My Country Too*) describes a grim sort of mobility, especially through the South, the car functioning as a protective capsule, at the same time triggering the protagonist's feelings of aggression.[37] Recently, this picture has been confirmed by two nonfiction studies by Mia Bay and Gretchen Sorin, the former providing an excellent historical empirics of Black car culture, the latter using her analysis as a celebration of Black 'freedom of mobility,' culminating in the civil rights movement's successful bus boycotts (and further popularized through a documentary).[38]

Very few *poor whites* figure as drivers or passengers in this phase's autopoetic universe, and if so, they do not write, but are written about, for instance in Harry Crews's 'southern grotesque' novels such as *Car*

(1972), where the utter mutual dependence on automotive mobility ("unlikable but necessary") is thematized through the eating of a car. The owner is 'owned' by the car-as-God, its consumption a sacrament.[39]

Whereas minorities and marginals celebrated the car's adventurousness as a road to emancipation, Crews's critical car stance reflected *mainstream automotive culture*. The iconic literary representative of this culture is John Updike's 'middle-American' hero Harry Angstrom in his quasi-historical Rabbit tetralogy. In the first volume, *Rabbit, Run* (1960), as well as in its sequel, *Rabbit Redux* (1971), fear is everywhere: fear of women, of the Bomb, of the Russians, of the dark (as a kid), of Blacks, of the law. A comparison with that other American automotive 'everyman,' Sinclair Lewis's equally mainstream prewar Babbitt (see Chapter 2), suggests that the car had now lost its role of provider of pleasure even during commuting, as Rabbit brings the *family car* back to his wife after leaving her, and after a failed escape by car (he returns after reaching West Virginia). Updike's narrative rhymes with that of John Keats, who in his nonfictional *Insolent Chariot* (1958), as one of the many critics of this period on the lost romance of prewar car driving, speaks of the 'terror or boredom' of driving along the newly built freeways. The imperative in Updike's title makes the novel a moralistic tract, as if the author himself incites his protagonist to move, but he won't, or can't.

The centrality of the car in Flannery O'Connor's *Wise Blood* (1952) reveals that also in her Southern Gothic universe the car has become ubiquitous, not as an adventure machine, but as a utilitarian tool. Even John Steinbeck's truck-cum-mobile-home in his travelogue *Travels with Charley* (1962) not only enables a flight from the world (the freeways prevent the author from seeing anything outside of his vehicle), but also from his 'Self,' contrary to the tendency among literary scholars to see literature as one great Quest of the Self. It is the freeway world that, 'flattening' the adventure, sets postwar autopoetic mobility apart from the interbellum culture.

This critical car stance in mainstream postwar literature begs for an explanation, which must be rather complex considering the giant immediate-postwar bacchanal of eroticized car adventure construed around the hot rod and drag racing subcultures in mainstream popular culture. But for this, we first have to make an excursion into *popular culture*: the number of American 'car songs' peaked in the first half of the 1960s at fifteen hundred, and not only celebrated affinities with car driving (Canned Heat's 1968 "On the Road Again," where the cadence of the music mimics the hammering of the rubber tires on the edges of the concrete slabs of the freeway), but also the pervasive use of the car and its components as metaphors ("I've got an engine / under my hood," Rod Stewart in "Let Me Be Your Car," [1974]). Or consider "white trash working-class male" and former truck driver Elvis Presley's over-conspicuous consumption to buy fourteen Cadillacs for his friends (and one bystander) in one day in 1975.[40]

The explanation of this split in automotive culture contains a mixture of class and lifestyle arguments. On the one hand we see a *youth culture* emerge, originally identified as *the beats*, with their bricolage of 'trips,' from hitchhiking to Benzedrine (amphetamine in the form of a bronchial inhaler), as an avant-garde away from the prewar automotive adventure as a driver's experience. Black beat poet LeRoi Jones called them "Americans no character in a John Updike novel would be happy to meet. [They] are . . . the last romantics of our age," quoted in the best analysis of beat mobility, written by Katie Mills.[41] The biography of Jack Kerouac and his *On the Road* (1957) illustrate how certain forms of countercultural subversive mobility became mainstream through a process of gradual absorption for which Marshall McLuhan coined the phrase 'repressive tolerance.' *On the Road* describes a journey in an over-the-top adventure machine driven by the protagonist's friend, from the perspective of the passenger. The book shows a paradoxical combination of a violent machismo and automotive narcissism, but at the same time this adventure has a rather abstract, near-existentialist character (Norman Mailer called the hipster "the American existentialist"), as it does not describe touring in the classical sense but as a 'pure,' nearly abstract enjoyment of the 'freedom' and detachment of the 'trip' and its 'speed,' homological to its "playful use of language as pure sound." In the beat culture there was no "close personal sensation with the car," in Deborah Clarke's analysis, not only in the US: Italian writer Luciano Bianciardi has one of his protagonists sigh, "traveling serves no purpose nowadays, and you don't learn anything from it."[42] Another characteristic of the beat culture is its nomadism, but this nomadism always has a destination: home. In a polemic with mobility scholar Tim Cresswell, Simon Rycroft concluded that "the beat experience was a sedentary one" (because of the destinations of their cross-country trips); yes, but this sedentarism was highly mobile, as the car trips were like sitting in a room on wheels.[43]

Last, beat mobility is characterized by multimediality, expressed by a plethora of 'affinities' between car driving and the production and reception experiences of other media, from film and music to television. Kerouac's and his automotive friend Cassady's 'spontaneous writing' rhymes with the latter's driving style, whereas avant-garde classical music (popular among several pop and rock artists) evokes the rapidly changing eye perspectives and in general the freeway experience of car driving. Television made the beats into mainstream celebrities: by abstractifying their mobility, and embedding it in a vaguely (no longer 'subliminal') ironic attitude, they prepared the final absorption of their culture into hegemonic culture, at a moment when automobilism, according to economic historian Walt Rostow, had "lost . . . the capacity to drive forward American growth."[44]

On the other hand (next to the beat culture), from the 1950s a working-class biker culture was revived, which gradually seemed to take over the adventurousness of the car, at least in popular culture. But even Marlon Brando's riding in *The Wild One* (1953) is "almost meditative," according to Ted Bishop. And Katie Mills found the rebellion of Billy and Wyatt (Dennis Hopper and Peter Fonda) in *Easy Rider* (1969) empty.[45] More scholarship is needed here to determine the actual adventurousness of biker culture, but film students have likewise analyzed contemporary Hollywood products (especially the American New Wave, inspired by European examples and headed by directors such as Sam Peckinpah and Steven Spielberg), in which the car acquired agency, but at the same time became abstract: driving is presented as an "enigmatic ritual," a "composition of pure movement" (in David Laderman's analysis of *Two Lane Blacktop* [1971, dir. Monte Hellman]), just as we observed in the case of Kerouac's *On the Road*. Or if the spatial aspects of adventurousness are celebrated, the driving is "aimless" (like Marnie's drifting between cities in Alfred Hitchcock's *Vertigo* [1958]).[46] Some road movies of this period made the adventure machine invisible (driving as an abstract, aimless practice), while others jumped into a flight forward, as a carnivalesque practice (see the iconic car chase in Peter Yates's *Bullitt* [1968] or the terrorizing of pedestrians at high speed in Stanley Kubrick's *A Clockwork Orange* [1971]). Whatever the angle, Hollywood scenario writers and directors sprinkled their irony around.

Such irony is absent in another working-class mobility culture created by Chicanos in Southern California in the late 1960s, which historian David Gartman described as a "spin-off movement of hot rodding" going back even to the 1930s when Mexicanos in Los Angeles put cement bags in their trunks to lower their cars.[47] But this playful but subversive form of the tinkering tradition within the automotive adventure (of adding colorful 'murals' to the lowered car body and inventing and mounting devices to lower the car while cruising through city streets) can better be seen as a heritage from the feminized forms of very early car adventure (such as the gymkhanas and flower contests, see Chapter 1, or the workers' parades in 1 May celebrations in Chapter 2) and the fashionable display of the Chicano zoot suiters with their outlandish trousers and colorful shirts with pearl buttons. Such 'overdecoration' played with what Mikhail Bakhtin has called the carnivalesque, as 'a means of transgressing binary oppositions, particularly that between high and low.'

[[In his dissertation (1940, published 1965, and in English translation in 1968) on the French early Renaissance Franciscan François Rabelais, Mikhail Mikhailovich Bakhtin (1895–1975), a literary theorist and philosopher in the Soviet Union, showed that medieval European folk humor created a "second life outside officialdom" as a temporal "liberation from all that is utilitarian."[48]

Bakhtin's theory is useful to help understand the collective actions of motorists against the grain. In our interpretation, the classic *carnivalesque mobility* is produced by the "mob," the seemingly unorganized group on the 'low' side of society who gave 'mobility' its first historical meaning, "obscene, lascivious, and scandalous" in their "unlawful games and interludes, drunkenness, etc."[49] The historical carnival may be seen as the birthplace of irony, as the folk form of irony, because in the reversal of social hierarchies the 'low' sets the 'high' apart, without excluding them in the festivities. In other words, Bakhtin provides a historical grounding of the ironic experience that goes far beyond the current philosophical father of irony, Søren Kierkegaard: originally, irony, in its coarse carnivalesque form, was of the people, and then it was captured by the middle class (as Kierkegaard sensed), at first by its cultural avant-garde, then, after World War II, invading the entire Western culture, through its mass media. Thus, the carnivalesque is the corporeal irony of the 'folk,' and as such, paradoxically, it is *post-ironic* (see the next chapter) in its refusal to use laughter as a distancing tool (except for briefly setting itself apart from the ruling class).]]

In other words, such mobility carnivals are the popular-cultural equivalent of middle-class irony: it is the ostentatious flight forward, the distanciation of the Other through noise or exuberance rather than through doublespeak. The 'pimped' Cadillac in the Black automotive subculture belongs to this carnivalesque part of the mobility culture, used as "symbolic affronts to the power of whites," often an object of a collective use culture (as a survey from 1959 revealed).[50]

Although mass motorization in Europe developed along similar lines, there were also some differences that made European postwar mobility into a model on its own, at least in its Western 'core.' Roadbuilding initially got underway through the Marshall Plan (1948–51), but soon two different consumption cultures developed, a German *bürgerlich* (petty bourgeois) one, somewhat skeptical toward the exuberance of the US, and a Fordist model followed by Sweden, France, Italy, and the Netherlands. 'Americanization' later would have these cultures converge, but the differences in personal prosperity between the 'core' and a southern and eastern 'periphery' remained large: a factor of 3 to 4 (and in some cases, such as Albania and Romania, even 6) in GDP per capita.[51] A new generation of scholars recently qualified Germany's 'Americanization' by pointing at its unique consumption culture (compared to the US) expressed, among other things, in an extensive use of public transport, not only by the poor, but also by the middle class.[52] Similar tendencies can be observed in other European countries.

Germany counted around one-fifth of its workers as car owners in 1962 (increasing to two-thirds a decade later), motorizing on its true 'national family vehicle,' the Volkswagen 'Beetle.' European 'extravagance' à la the US manifested itself in Italy and France in the small car, but also in the proliferation of the motorcycle, and the moped, especially among the youth. The Italian Piaggio increased the production of its Vespa scooter (as "a cure against communism") from 700,000 in 1956 to 4.3 million eight years later, dethroning the British motorcycle industry from its dominant position in the world. But the Vespa also functioned in a subversive context: the mods (as British low-class dandies were called) 'masculinized' the two-wheeler during their "scooter charge" on Buckingham Palace in November 1966, or in the Battle of Brighton against the 'lower' working-class rockers.[53] In the Netherlands, working-class youth of Indonesian descent preferred the Austrian Puch moped (they regularly took the engine apart, just for the fun of it), their girlfriends on the French Solex or (when this two-wheeler acquired the reputation of being an old ladies moped) the Mobylette, the same moped that in Turkey was used by "rowdy *mobilet* gangs," testifying to the fact that the 'script' of artifacts, including vehicles, is extremely flexible indeed (if not a useless category).[54] In Sweden, working-class youth could afford to buy American gas guzzlers, constructing a collective culture around them, equipping them with foxtails and other proletarian paraphernalia as the mods did with their scooters. Whereas Finnish working-class youth and Italian *teppisti* (hooligans) formed their own specific subversive car cultures, Amsterdam's anarchist youth developed a free and collective 'white bicycle plan' as an alternative to the car. It was through such groups that mainstream car culture was first resisted, until then accepted by beats and countercultural youngsters as a matter of course. And it was through such youthful groups that parts of a proletarian or plebeian lifestyle and values ("roughness, excitement, chance-taking, indulgence, 'conning,' autonomy and hardness") were absorbed within the middle class.[55]

Like in the US, international tourism in Europe was stimulated by the construction of a freeway network. A crucial part of the Marshall aid was 'travel development,' enjoyed not least by American tourists who spent twice America's foreign aid budget in Europe in 1955. They formed only a miniscule part of a worldwide trend of international tourism, which grew from 25 million in 1950 to about 156 million in 1970. In Europe, after tourism champion France, it was fascist Spain that profited the most from these massive migratory flows, exploding from 1.5 million (1955) to a staggering 40 million in 1978, more than its 37 million inhabitants. In the other direction a partly illegal bus line network emerged from the late 1950s, transporting Portuguese and Spanish 'guest workers' to the northwest, until it was replaced by the car and the plane, its remaining lines merging into the Eurobus network, the European Greyhound. From the 1960s a worldwide flow of two million men and women migrated to richer countries, followed in size by flows of refugees, both phenomena resulting in a "feminization of migration."[56]

Japan's increasing car production kept the American share in check, not only in Europe but in the entire world: the share of American cars on the world market diminished from three-quarters in 1955 to one-third in 1973, whereas 'Europe' took a more or less constant share of one-quarter.

If the content of European autopoetic novels is any indication of the continent's 'typical' car culture, it may be proposed that this culture was somewhat more adventurous than the 'flattened' one in the US, but also perhaps a bit more status sensitive. It is, for instance, hard to find an American mainstream equivalent of Françoise Sagan's car addiction: after her heavy car accident in 1957, she bought a Ferrari (in 1966), terrorizing nightly Paris at 200 km/h.[57]

On the other hand, we find equally 'flattened out' adventures in contemporary European autopoetics, from Martin Walser's *Halbzeit* (Halftime, 1960) to Muriel Spark's *nouveau roman* and female Gothic novel *The Driver's Seat* (1970), but we need more analyses of immediate-postwar European autopoetic literature, including those of the East and South, to better enable us to compare European and American autopoetic universes. The same applies to the autopoetic movies of the European New Wave, where perhaps a more open criticism of mass motorization is to be found in films such as Jean-Luc Godard's *Weekend* (1967) or Luigi Commencini's *Ingorgio* (Jam, 1978), next to the celebration of adventure in Michelangelo Antonioni's *L'avventura* (1960) or Federico Fellini's *La Dolce Vita* (The Pleasant Life, 1960).

Summarizing, in the immediate postwar period of transition it was in the West where the car not only continued its prewar quantitative and cultural dominance, but did so through an exuberant indulgence in travel by car, especially so in the United States. To enable this democratization beyond the adventure machine's original inventor, the elite and the middle class, the beast lost several of its adventurous feathers. Paradoxically, new avant-gardes emerged, celebrating the car's adventurous traits in their emancipatory ambitions, traits that returned in an inflated, carnivalesque form within popular culture, from songs to movies, thus increasing the further fragmentation of international automobilism.

It was this kaleidoscopic amalgam of car cultures that formed the basis of the globalization of the 'car society' during the second half of the twentieth century, but this took place in a special way.[58] Instead of an increasing car fleet demanding an infrastructure, it was the other way around: the proliferation of the car society over the globe was primarily *systemic*, through the export of road networks, including the expertise to plan them, the funds to finance them, and the machines to build them. The way this was accomplished was through 'development,' aimed at creating new interdependencies between 'core' and 'periphery' in a globalizing world.

The aggressive, American form of road promotion was first tested on Europe-as-semi-periphery, through Marshall funds. With the hegemony in transnational road engineering already before the war having shifted from Europe to the United States, the civil-servant-dominated roadbuilding association PIARC saw a rival organization emerge, the International Road Federation (IRF). Its management was recruited from the car industry, who published a glossy magazine *Road International* and appeared to be interested especially in spectacular, large-scale roadbuilding projects, as outright roadbuilding propaganda. The IRF acquired, through maneuvering by the American government, a consultancy status at the United Nations, but under the pressure of the emerging Cold War both associations' ideologies gradually converged, resulting in common conferences and other forms of cooperation, such as the recruiting of members to receive courses in traffic engineering at Yale University, an institute from which the roadbuilders' bible, O. K. Norman's *Highway Capacity Manual*, was also spread (even into the Soviet Union where it was translated in the mid-1960s).[59] Together with the bold construction of the gigantic Interstate Highway System, the US became an example of how to plan and build limited-access freeways, not only through the founding of 'quasi-public corporations' (such as the Tennessee Valley Authority during the New Deal [1933]), but also by what British critics call 'predict and provide' procedures, the practice of constructing roads on the basis of (often rather shaky) estimates of future car use.

Thus, in Germany, the first road plan was never brought into parliament, enabling a golden era during the 1960s and early 1970s of designing a nationwide freeway network, built upon a substrate of prewar Nazi *autobahnen*, superposed upon an independent system of state roads (*Bundesstrassen*) and coordinated by a central planning division within the Ministry of Transport. In Switzerland, parliament disempowered itself when it started to plan a system that not only was meant to alleviate urban congestion but also to keep the country connected to the emerging European network. In France, with a weak and young regional governance structure, some regions in the south only got connected to the national grid in the 1990s. Even Spain began constructing a network in the mid-1960s. All these national networks formed, from below as it were, at the European level, a so-called E network, complemented in the East by the Yugoslav north–south highway.[60] The decade around 1970 can be called a golden era of European freeway building. In the US freeway building peaked in 1966, after which a period of resistance set in, soon followed by Europe, which led to the end of the 'planning euphoria' and to the emergence of what specialists called 'comprehensive planning,' in which public transport was also integrated.

The spread of the roadbuilding-as-development mantra over the globe was inspired by 'liberal anti-communism,' often attributed to economic historian Walter Rostow's *Non-Communist Manifesto* (as the subtitle ran of his *Stages of Economic Growth*, 1960), in which 'automobile dependence' was considered to be the highest stage, and other nations would 'catch up' after their 'take-off' into modernization. Unhistorical and deterministic, and also chauvinist (modernization is "a genuinely American phenomenon"), his linear narrative observed how all Western nations and Japan "behave[ed] in a remarkably 'American' manner."[61] In the West this diffusionist ideology, spread by myriad historians, social scientists, and traffic engineers, influenced decisively the so-called coordination crisis in prioritizing road transport over railways, and was in fact fully based on expectations, on the belief in "modernity [as] a set of promissory notes," in Björn Wittrock's words.[62] For instance, in 1964 Wilfred Owen wrote a programmatic monograph on *Strategy for Mobility* for the Brookings Institution (funded by the US Agency for International Development, USAID) in which he emphasized the importance of investments in roads rather than rails (even when he conceded that in his research a 'highway' truck appeared to be 4.5 times more expensive per ton-mile than the railway), at a moment when half of the World Bank's international funding was still on rails, and only one-third on roads.[63] But development was much more than an economist's practice. Modernization was supposed to create a 'mobile personality,' as David Lerner identified in his study of the Middle East: an individual with a "rationalist and positivist spirit," capable of identifying with newness and change, against whom Islam would be "absolutely defenseless."[64]

When Rostow left MIT's Center for International Studies to become President Kennedy's national security adviser, he suggested to make the 1960s into an 'economic development decade,' especially aimed at strengthening Latin America's emerging national middle classes, often neglected by the local Communists. Also, a shift in the balance in US foreign policy occurred, adding coercion (coups in Peru, Ecuador, the Dominican Republic, Honduras, Guatemala, and Argentina) to the 'soft power' of the Peace Corps and its 70,000 volunteers, the latter supported by new social science disciplines, introduced as 'area studies.'

And although there was (and is) no consensus among transport experts about the relationship between roadbuilding and economic growth, the belief in the magic of transport investments remained unperturbed, although one of the founders of the theory, Columbia University economy professor Albert Hirschman, in order to save this theory, had to make a distinction between 'development by shortage' and 'development by excess,' opting for the former and implying that many road investment programs, when following the latter, were realized ahead of demand, as expectations cast in concrete.

By mid-1963, sixty-four countries (one-third of them in Asia, and one-quarter each in Central/South America and in Europe) had received from the World Bank $7 billion for eight hundred projects, one-third of the total spent on mobility infrastructures. By then the bank's lending policy had shifted to a 'basic human needs approach' and during the presidency of former General Motors CEO and 'Cold War warrior' Robert McNamara (1968–81) it shifted further to poverty relief, bringing into focus the construction of agricultural 'low-cost' roads. That happened at a moment, the end of the 1960s, when "the old doctrines of development . . . fell away with remarkable speed."[65]

One of the triggers of this 'end of the planning euphoria' was the founding of the United Nations Conference for Trade and Development (UNCTAD, 1962), which during its first conference in Geneva in 1964 had Ernesto 'Che' Guevara on behalf of the Cuban delegation warn against the imperialism of the IMF and World Bank.[66] Another was the emergence of an alternative theory of development within the Comisión Económica para América Latina (later extended with 'y el Caribe'), CEPAL, of the United Nations (also known under its English abbreviation ECLA), where economist Paul Baran, in a historical analysis of the failure of Eurocentric modernization liberalism, coined the phrase 'development of underdevelopment.' The theory of the *dependentistas* expanded beyond Latin America through the 'world system analysis' coined by Immanuel Wallerstein, which in its turn provoked a revival of developmentalism by a younger generation of social scientists dealing with the modernization of the 'Asian Tigers.'[67] But the main result of the new theories was that they brought agency back into the core–periphery equation, on the side of the periphery. By the end of the period covered by this chapter, the ILO (International Labor Organization) in Geneva also broke with modernization theory, particularly because of the growing recognition of the importance of the 'informal sector' and its (often 'illegal') mobility.

But empirics are often stronger than theories: whether developmentalist or *dependentista*, development-through-roadbuilding continued to be rolled out over the entire globe. Apart from highly symbolic and megalomaniac projects such as the Pan-American Highway, heavily supported by the IRF, all *Latin American countries* started ambitious road programs at a moment, the end of the 1960s, when nine of them had their own automotive assembly industries. In Mexico a destructive competition between rail and road started already during World War II. By 1950 half of all freight there was carried in trucks gathered in a state-sanctioned Alianza de Camiones. But as late as 1970, still only one-fifth of all trips in Mexico City were done by car, which competed for road space with a diversity of (semi)public transport such as taxis, *peseros*, combis, micros, electric trolleys, and the new Metro. Only one-fourth of all *poblados* were connected by roads. Yet several Latin American countries can be considered roadbuilding laboratories, in the Mexican case through the issuing of 'road bonds' purchased by American investors, promoted by the IRF and its local monthly *Más Caminos* (More Roads). During the 1960s, Mexico's road network increased from 45,000 to 71,000 km, at a moment (1965) when there were 1.2 million motor vehicles in the country. An even more outspoken roadbuilding laboratory (a "territory of experimentation," according to historian José Vielma-Cabruja)[68] was Venezuela, which realized a part of the Pan-American at an altitude of 3,000 feet.

In many cases roadbuilding took place well before mass motorization, with some exceptions, such as the *vida digna* campaign in Peronist Argentina, which rejected car hedonism (and could do so, one is inclined to conclude, because of the partial disempowerment of the middle class). In 1964 four million motor vehicles were counted on the continent (1.7 million in Brazil, 1.4 million in Argentina, 1.1 million in Mexico [all with an extensive car industry], but only 2,200 in French Guiana). Brazil, Argentina, Chile, and Colombia doubled the length of their road network during the period covered by this chapter, whereas the motor vehicle fleet of the entire continent increased by a factor of 4 (for cars) and 5 (for trucks).

The IRF's influence (through the organization of 'regional road congresses' for instance, but also through its propaganda in *Road Inter-*

national) was also crucial in the construction of the postcolonial road network in *Africa*, its development analyzed as no other by anthropologists. With the collapse of colonial regimes in Africa, migrations and reverse migrations of colonial personnel, settlers, and their descendants took place, while independence wars and post-independent factional wars created flows of refugees.[69] 'Modernization' took place in many shapes and fashions. For instance, in Kenyan Nairobi, a colonial master plan in 1948 refurbished the former garden city into a functionalist separation of living, working, leisure, and movement, thus perpetuating the city's racial segregation, because functionalism is dependent on mobility and "Nairobi's road network, structured around radial axes, favored control over facilitating movement," in Robert Heinze's analysis, which led to the "schizophrenic heritage—demanding movement while impeding it,"[70] a nice example of what we have come to call mobility Orientalism.

> Mobility Orientalism is tributary to Edward Said's *Orientalism*, the seminal attack on the racist Western gaze on the 'East,' but directly inspired by Daniel Vukovich's analysis of "Sinological-Orientalism and its production of a textual 'China' [that helps] constitute the identity of 'Self' of the West," not by 'othering' the Chinese but by declaring that "the West, and the U.S. in particular, is what China is *not* but which the latter must, and will someday become . . . China is seen as in the process of haltingly but inevitably becoming-the-same as 'us': 'open,' liberal, modern, free." This is part of a more general strategy of, for instance, "European colonial anthropology [that] tended to construct non-European others as objects of lack," a reflex we have also seen, earlier in this chapter, appear in Soviet mobility historiography.[71] From this perspective, Rostow's *Stages of Economic Growth* is mobility Orientalism's manifesto.
>
> Mobility Orientalism is the textual construction of the 'mobile other' in the Global South (and, as we saw, in Eastern Europe) as exotic and backward, but at the same time it encompasses the proposed solutions aimed (consciously or not) at making its mobility the same as 'ours.' In this and the next chapter we will encounter many examples of this 'modernization' reflex, enhanced and enabled by "'self-Orientalizing' procedures" by the local elites and their stakeholders.[72]

The indigenous answer to this form of top-down modernization was often 'illegal,' for instance in the case of Nairobi, where in the late 1950s "pirate taxis" appeared on the streets, as a response to the colonial "hire car system" that privileged European and Asian entrepreneurs. Such taxis soon functioned in a shared-transport setting, "focus[ing] on specific routes, particularly during rush hours." Later, converted into station wagons, such vehicles became known as *matatus* (see Chapter 4).[73]

Apart from the very north and south of the continent, the most advanced among African countries in the race toward modernity was Ghana (former Gold Coast) with its population constantly on the move (half of its workforce in 1960 consisted of 'mobile workers'), an important part of which consisted of women whose freight transportation by head porterage on a market day (moving more than 80 percent of all commodities [an impressive 65.5 tons]) was translated into sophisticated flowcharts by anthropologist Peter Gould.[74] In a comparative study of Ghana and Nigeria, Africanist Gina Porter showed how the former country's rural markets collapsed and the latter country profited from its oil reserves during the oil price hikes of the 1970s (see the next chapter), resulting not only in the sudden growth of the motorcycle and car fleet, but also of the number of entrepreneurs who started taxi and mammy wagon services. The latter, often trucks repurposed for passenger transport, became a part of a growing 'informal economy' of 'petty capitalism,' a form of value creation in a 'dualist' society (a small development sector and a large subsistence sector involving the rural masses), a term coined by economist Jan Boeke and brought into the mainstream by economists related to the ILO.[75]

The existence of large differences in motorization patterns between countries begs for explanations way beyond economics (which emphasizes the relationship between car density and income), but it cannot be denied that the differences between Africa's and Latin America's motorization (as well as between countries within these continents) are to a large extent explainable by the existence (or not) of a prosperous middle class. This is, however, not only a question of economics, but also of culture: there has to be developed a *desire* to motorize, to spend money on a motor vehicle above other options. A survey by the University of Ibadan in 1969 of Nigerians' travel motives found the 'family motive' to be dominant: men traveled to buy, women to sell, but the latter dominated in the flows. Hausa migrants dominated truck and bus transport, its analysis by Adrian Peace not only showing its fundamental collective character, but also the adventurous, if not carnivalesque 'ruthless' driving culture of Hausa bus drivers.[76] The layeredness of Nigerian mobility was not only caused by the superposition of 'new' and 'old' transport technologies, but also by the 'mobility' between a rural and urban identity, as can be found in Cyprian Ekwensi's highly successful novel *Jagua Nana* (1961) in which the protagonist departs from a motor park in Lagos, sitting on the truck bed of a mammy wagon, and is brought to her destination in the countryside by a bicycle taxi.

One *alternative development* model has recently been charted by STS professor Joshua Grace, using oral history and participant observation by joining a garage, among other things, enabling us to compare it with the Soviet case discussed earlier in this chapter.[77] Like the Soviet Union, considered a 'roadless' country, colonial Tanganyika already early on knew a culture of roadside repair in 'muted garages' (*gereji bubu* in Kiswahili) where a masculine tradition of learning-by-doing was developed, just as in many other 'peripheral' countries in the Global South. By emphasizing "sensory skills" and "sensual knowledge . . . outside of European technical authority," Grace mobilizes recent technology historians' insights into the history of repair and maintenance to depict an alternative route to car technology expertise. After independence (1961), and differently from the Soviet Union, Tanzania's government tried to capitalize on this tradition by setting up a network of regional workshops, from which mobile teams of 'state mechanics' traveled the country in Land Rovers to tackle the most complex repairs. But the tradition of informal workshops continued as well: the government considered the 'home garages' a threat to the building of a socialist society, which in the case of Tanzania took shape through an anti-urban ideology focused on farming in rural communities. There, *mikoani* (countryside mechanics) criticized their urban colleagues as practicing a 'take-off and put-on' policy, whereas they had to be creative without much access to spare parts and machines.

When, in 1967, Tanzania's government officially turned to socialism with the Arusha Declaration, the country's roadbuilding program emphasized the low-cost 'feeder roads' next to a highway to Zambia (built with foreign aid). Instead of cities and cars, Grace concluded, "hoes, villages, ox plows, and walking . . . took center stage" during the 1960s.[78] Whereas villages could buy trucks, urban mobility socialism developed by *reducing* private car ownership, while from the early 1970s onward illegal minibuses (*thumi-thumi*) were used for semipublic transport, which were legalized in the next phase (see Chapter 4) and then were called *daladala*. It was a masculine form of mobility socialism, which made the recorded life of 'auto-mobile' Hawa Ramadhani all the more remarkable: her exceptional career as a bus driver (costing her her marriage) lives on in a children's book, *Hawa the Bus Driver* (1987), read in Tanzania's primary schools.[79]

In the early Tanzanian story the IRF does not seem to figure. Nor is the influence of the IRF's road bias on the United Nations well researched, although it is remarkable that the UN's regional economic commissions also began to formulate grand visions, such as the ECA (Economic Commission for Africa), which initiated the Trans-African

Highway network from West to East, a suggestion made by a Japanese economic mission to Africa. Mitsubishi Corporation was willing to build the road, and the Japanese government to give aid.

From the early 1960s Japan began to export capital to its neighbors, mostly through its membership of the South East Asia Treaty Organization (SEATO). In Asia, too, the ECAFE worked on an Asian Highway, after planners within the commission initially considered transport not a priority, as they believed, in the analysis of historian David Wightman (who wrote an invited history of the ESCAP), that "transport existed primarily to serve the trading interests of the imperial power."[80] The network encompassed nearly 63,000 km of roads through fourteen Asian countries. Japan became a full ESCAP member in 1954 and competed with the Western members (Australia, France, the Netherlands, the Soviet Union, the UK, and US) in the export of transport expertise, among other things through a tour of Japan National Railways. On the other hand, within ESCAP knowledge production about low-cost roads also occurred, for instance in the founding, in 1959, of a SEATO Graduate School of Engineering in Bangkok. For many Asian countries, roadbuilding activities during the period covered by this chapter still have to be charted.

Summarizing, in this transitional period of world mobility history (which for Western observers may have been experienced as the pinnacle of mobility modernity), the globalization of a Western car society model took place through the lucrative export of road network building, stimulated by the World Bank and IMF. It showed that decolonization, development, and globalization are closely intertwined, with automobilism as one of the decisive areas.

Whereas in the West, especially in the US, automotive culture at the same time became excessive and 'flattened' (routinized), emancipatory subcultures of women, Blacks, and other 'marginals' kept an adventurousness alive, picked up by a popular culture that morphed it into a carnivalesque flight forward. One wonders how this fragmented culture was absorbed in the Global South, and how it was transformed at a moment when car cultures began to emerge there.

Although we will study this in the next chapter, covering the decades leading to the turn of the twenty-first century and beyond, at the closing of the post-World War II period until the energy crises of the 1970s, it is important to stress, on the evidence brought forward in this chapter, that 'development-through-roadbuilding' was not a unidirectional process. Not only did the majority of the emerging middle classes motorize on the two-wheeler, in some (Asian) countries urbanites developed their own rickshaw culture, whereas the Soviet Union opted for the railways as the basis of its modernization and initially even intended to refrain from the car in private ownership, followed by Tanzania. Last, we see in many countries of the Global South a form of informal semipublic transport emerge, making use of the car and the van, but deploying them in a collective culture.

And yet, local elites (who had embraced roadbuilding in the first place) now grabbed their chance to motorize à la the West. Why did they do so, and how did they decide to follow the West's example? And why did their subaltern 'others' not follow their example, or at least less so? Was it the Western consultants and the World Bank representatives who convinced the local elites? And if so, were their arguments drenched in 'mobility Orientalism?' Were they won over by touristic and business visits to the West, by stories coming back from their compatriots who had migrated there, by Hollywood movies, by reading the novels we have discussed here? We do not know. We need more detailed analyses of local mobilities. The next chapter can only provide some preliminary basics for such analyses.

Notes

1. J. G. Ballard, *Miracles of Life: Shanghai to Shepperton; An Autobiography* (London: Fourth Estate, 2008), 121.
2. See for the following Gijs Mom, *Globalizing Automobilism: Exuberance and the Emergence of Layered Mobility, 1900–1980* (New York: Berghahn, 2020), 205–9, 250–51, 394–415, 470–81.
3. M. William Steele, "A Modern Infrastructure of Manchukuo: Where Did All the Cement Come from?" (paper presented at the East Asian Environmental History Conference, Kagawa University, Japan, 22–25 October 2015), 13.
4. For an overview of such memoirs see Shuyu Kong, "Swan and Spider Eater in Problematic Memoirs of Cultural Revolution," *Positions* 7, no. 1 (1999): 239–52.
5. David Edgerton, "Creole Technologies and Global Histories: Rethinking How Things Travel in Space and Time," *HOST* 1 (Summer 2007): 75–112.
6. Prema-Chandra Athukorala and C. Veeramani, "From Import Substitution to Integration into Global Production Networks: The Case of the Indian Automobile Industry," *Asian Development Review* 36, no. 2 (2019): 72–99, here 77.
7. Holland Hunter, *Soviet Transport Experience: Its Lessons for Other Countries* (Washington, DC: Brookings Institution, 1968).
8. Elvira Khairullina and Luis Santos y Ganges, "Tram, Trolleybus and Bus Services in Eastern-European Socialist Urban Planning: Case Studies of Magdeburg, Ostrava and Oryol (1950s and 1960s)," *Journal of Transport History* 42, no. 1 (2021): 26–57, here 30, 33, 36, 47n72 ("reasonable"), 48 (transition period).
9. For instance by Siegelbaum in his introduction to the edited volume *The Socialist Car: Automobility in the Eastern Bloc* (Ithaca, NY: Cornell University Press, 2011), 1–13, here 2, where he sees the car bringing its owners "a little closer to an imagined West." Overemphasis of modernization theory was already observed by Michael Burawoy, "Review: The End of Sovietology and the Renais-

sance of Modernization Theory," *Contemporary Sociology* 21, no. 6 (November 1992): 774–85. Kornai is referred to in Katherine Verdery, *What Was Socialism, and What Comes Next?* (Princeton, NJ: Princeton University Press, 1996), 21. 'Lacking': Jeremy Morris and Abel Polese, "Introduction: Informality—Enduring Practices, Entwined Livelihoods," in Jeremy Morris and Abel Polese (eds.), *The Informal Post-Socialist Economy: Embedded Practices and Livelihoods* (London: Routledge, 2014), 1–18, here 7.

10. Luminita Gatejel, *Warten, hoffen und endlich fahren: Auto und Sozialismus in der Sowjetunion, in Rumänien und der DDR (1956–1989/91)* (Frankfurt: Campus, 2014), 109; Maria R. Zezina, "The Introduction of Motor Vehicles on a Mass Scale in the USSR," in Manfred Grieger, Ulrike Gutzmann, and Dirk Schlinkert (ed.), *Towards Mobility: Varieties of Automobilism in East and West* (Wolfsburg: Volkswagen AG, 2009), 43–53, here 50; *Gazeta*: Luminita Gatejel, "The Wheels of Desire: Automobility Discourses in the Soviet Union," in ibid., 31–41, here 37; US: Corinna Kuhr-Korolev and Dirk Schlinkert, "Introduction," in ibid., 9–20, here 15.
11. György Péteri, "Alternative Modernity? Everyday Practices of Elite Mobility in Communist Hungary, 1956–1980," in Siegelbaum (ed.), *The Socialist Car*, 47–68, here 47, 53, 59, 66 ("feasible": 49); György Péteri, "Streetcars of Desire: Cars and Automobilism in Communist Hungary (1958–70)," *Social History* 34, no. 1 (February 2009): 1–28.
12. Martin Crouch, "Problems of Soviet Urban Transport," *Soviet Studies* 31, no. 2 (April 1979): 231–56, here 231.
13. This is consistent with Lewis H. Siegelbaum, "On the Side: Car Culture in the USSR, 1960s–1980s," *Technology and Culture* 50, no. 1 (January 2009): 1–23: apart from some articles in the car enthusiasts' magazine *Za rulem* from the second half of the 1960s, most of his sources date from 1975 or later. Before 1964 only half (61,000 cars on average) of the total annual production of cars were "sold to individuals." Jukka Gronow and Sergei Zhuravlev, "Soviet Luxuries from Champagne to Private Cars," in David Crowley and Susan E. Reid (eds.), *Pleasures in Socialism: Leisure and Luxury in the Eastern Bloc* (Evanston, IL: Northwestern University Press, 2010), 121–46, here 134–35.
14. Luminita Gatejel, "Privat oder Staatlich? Automobile Konsumkultur in der Sowjetunion, der DDR und Rumänien," in Manuel Schramm (ed.), *Vergleich und Transfer in der Konsumgeschichte* (Leipzig: Leipziger Universitätsverlag, 2009), 16–32, here 31.
15. Luminita Gatejel, "The Common Heritage of the Socialist Car Culture," in Siegelbaum (ed.), *The Socialist Car*, 143–56, here 147, 155–56.
16. Gronow and Zhuravlev, "Soviet Luxuries from Champagne to Private Cars," 122–23, 126 (Table 4.1), 132 (Table 4.2); large cities: Siegelbaum, "On the Side," 7–8, 10.
17. Howard Dick and Peter J. Rimmer, *Cities, Transport and Communications: The Integration of Southeast Asia since 1850* (New York: Palgrave Macmillan, 2003), 71.
18. Peter J. Rimmer, *Rikisha to Rapid Transit: Urban Public Transport Systems and Policy in Southeast Asia* (Sydney: Pergamon Press, 1986).
19. Tim Edensor and Mark Jayne, "Introduction: Urban Theory beyond the West," in Tim Edensor and Mark Jayne (eds.), *Urban Theory beyond the West: A World of Cities* (London: Routledge, 2012), 1–27.
20. Giovanni Arrighi, *The Long Twentieth Century: Money, Power, and the Origins of Our Times* (London: Verso, 1994), 344.

21. M. William Steele, "The Making of a Bicycle Nation: Japan," *Transfers* 2, no. 2 (Summer 2012), 70–94, here 74–79 (quote on 77); thirty million: M. William Steele, "The Speedy Feet of the Nation: Bicycles and Everyday Mobility in Modern Japan," *Journal of Transport History* 31, no. 2 (December 2010): 182–209, here 200, 202.
22. Ezra F. Vogel, *Japan's New Middle Class: The Salary Man and His Family in a Tokyo Suburb* (Berkeley: University of California Press, 1971), 5, 277.
23. Mary McShane, Masaki Koshi, and Olof Lundin, "Public Policy toward the Automobile: A Comparative Look at Japan and Sweden," *Transportation Research Part A* 18A, no. 2 (1984): 97–109, here 99–100.
24. See for the following Mom, *Globalizing Automobilism*, 245–351.
25. Arrighi, *The Long Twentieth Century*, 277–87, 303–6, 342 (the 'chaos' quote [on 285] is from Franz Schurmann, 1974); Wall Street: Verena Kröss, "Profitability through Quantification: The World Bank and Rural Development in the 1970s," in Corinne R. Unger, Nicholas Ferns, Jack Loveridge, and Iris Borowy (eds.), *Perspectives on the History of Global Development* (Berlin: Walter de Gruyter, 2022), 183–204, here 185.
26. Quoted in Robert Braun and Richard Randell, *Post-Automobility Futures: Technology, Power, and Imaginaries* (Lanham, MD: Rowman and Littlefield, 2022), 99.
27. Grant Madsen, *Sovereign Soldiers: How the U.S. Military Transformed the Global Economy after World War II* (Philadelphia: University of Pennsylvania Press, 2018), 235; Truman: Richard F. Weingroff, "The Man Who Loved Roads," *Public Roads* 65, no. 6 (May/June 2002): 37–46.
28. Avner Offer, *The Challenge of Affluence: Self-Control and Well-Being in the United States and Britain since 1950* (Oxford: Oxford University Press, 2006), 206; Shelley Nickles, "More is Better: Mass Consumption, Gender, and Class Identity in Postwar America," *American Quarterly* 54, no. 4 (December 2002): 581–622, here 583.
29. Ibid., 598.
30. Adorno's *Minima Moralia* (1974) as quoted in David Gartman, "Three Ages of the Automobile: The Cultural Logics of the Car," *Theory, Culture & Society* 21. no. 4/5 (2004): 169–95, here 181.
31. Margaret Walsh, "Gendering Mobility: Women, Work and Automobility in the United States," *History* 93, no. 3 (July 2008): 376–95, here 395: "it was no longer possible to perceive the automobile . . . as a masculine technology." Ellen J. Gerland and Craig L. Davis, "Selling Detroit on Women: *Woman's Day* and Auto Advertising, 1964–82," *Journalism History* 38, no. 4 (Winter 2013): 209–20, here 213.
32. Cotten Seiler, "'So That We as a Race Might Have Something Authentic to Travel By': African American Automobility and Cold-War Liberalism," *American Quarterly* 58, no. 4 (December 2006): 1091–1117, here 1094.
33. Mia Bay, *Traveling Black: A Story of Race and Resistance* (Cambridge, MA: Harvard University Press, 2021), 289.
34. Rainer Schönhammer, *Das Leiden am Beifahren: Frauen und Männer auf dem Sitz rechts* (Göttingen: Vandenhoeck & Ruprecht, 1995).
35. Discussed in Mom, *Globalizing Automobilism*, 284–351.
36. Deborah Clarke, *Driving Women: Fiction and Automobile Culture in Twentieth-Century America* (Baltimore: Johns Hopkins University Press, 2007); Katherine Lawrie Mills, "Remapping the Road Story: Contemporary Narratives of Autonomy and Mobility in American Literature, Film, and Television," dissertation (Los

Angeles: University of Southern California, May 2000); Ronald Primeau, "From Ma Joad to Elizabeth Berg: Women on the Road in America," *Midamerica* xxvi (1999): 138–46, here 138.

37. Roger N. Casey, *Textual Vehicles: The Automobile in American Literature* (New York: Garland, 1997), 134.
38. Bay, *Traveling Black*; Gretchen Sorin, *Driving While Black; African American Travel and the Road to Civil Rights* (New York: Liveright, 2020).
39. Harry Crews, "Car: A Novel," in Harry Crews, *Classic Crews: A Harry Crews Reader* (New York: Touchstone/Simon & Schuster, 1995), 329–436.
40. Nancy Isenberg, *White Trash: The 400-Year Untold History of Class in America* (New York: Viking, 2016), 231.
41. Mills, "Remapping the Road Story," 37.
42. Ellis Amburn, *Subterranean Kerouac: The Hidden Life of Jack Kerouac* (New York: St. Martin's Press, 1998), 52 ("pure"); Norman Mailer, "The White Negro: Superficial Reflections on the Hipster," in *Advertisements for Myself* (Cambridge, MA: Harvard University Press, 1992 [1959]), 337–58, here 339; Luciano Bianciardi, *La Vita Agra or It's a Hard Life* (London: Hodder and Stoughton, 1965 [1962]), 185.
43. Simon Rycroft, "Changing Lanes: Textuality off and on the Road," *Transactions of the Institute of British Geographers* 21 (1996): 425–28, here 427; see also Tim Cresswell, "Mobility as Resistance: A Geographical Reading of Kerouac's 'On the Road,'" *Transactions of the Institute of British Geographers* 18 (1993): 249–63.
44. W. W. Rostow, *The Stages of Economic Growth: A Non-Communist Manifesto* (Cambridge: Cambridge University Press, 1971 [1960]), xv.
45. Ted Bishop, "*Tempo Giusto*: The Art of the Slow Ride," in Klaus Benesch (ed.), *Culture and Mobility* (Heidelberg: Universitätsverlag Winter, 2013), 81–88, here 83; Katie Mills, *The Road Story and the Rebel: Moving through Film, Fiction, and Television* (Carbondale: Southern Illinois University Press, 2006), 122.
46. David Laderman, *Driving Visions: Exploring the Road Movie* (Austin: University of Texas Press, 2002), 93–105; Kathleen McHugh, "Women in Traffic: L.A. Autobiography," *South Atlantic Quarterly* 97, no. 2 (Spring 1998): 391–412, here 393 (Hitchcock).
47. David Gartman, *Auto Opium: A Social History of American Automobile Design* (New York: Routledge, 1994), 171–72.
48. Mikhail Bakhtin, *Rabelais and His World* (Bloomington: Indiana University Press, 1984), 6, 276. See for the following Gijs Mom, *Pacific Automobilism: Adventure, Status and the Carnival of Mobility, 1975–2015* (New York: Berghahn, 2022), 9–14.
49. Peter Stallybrass and Allon White, *The Politics and Poetics of Transgression* (Ithaca, NY: Cornell University Press, 1986), 32–33.
50. Jeremy Packer, *Mobility without Mayhem: Safety, Cars, and Citizenship* (Durham, NC: Duke University Press, 2008), 200–5; Mary Pattillo-McCoy, *Black Picket Fences: Privilege and Peril among the Black Middle Class* (Chicago: University of Chicago Press, 1999), 147 ("affronts").
51. Josef W. Schödermeier, "Die Entwicklung und Zusammensetzung des Personenkraftwagenbestandes in Westeuropa (Eine retrospektive Analyse)," dissertation (Cologne: Universität Köln, 1961), 44.
52. For instance Jan L. Logemann, *Trams or Tailfins? Public and Private Prosperity in Postwar West Germany and the United States* (Chicago: University of Chicago Press, 2012).
53. Silvia Cassamagnaghi, "Vespa Piaggio: First Mass Motorization and Discovery of Leisure in Post-War Italy" (paper presented at the conference "Inventing Eu-

rope: Eurocores," European Science Foundation workshop, Amsterdam, 15–17 January 2009), 11–12; Dick Hebdige, *Hiding in the Light: On Images and Things* (London: Routledge, 1988), 84, 109.

54. Wim de Jong and Bas van Kleef, *De Puch-story en andere brommerverhalen* (Bussum: Thoth, 1994), 38, 51, 56, 119; Bahar Emgin, "Rowdy Gangs on the Road: The Culture of Mobilet in Turkey," in Tevfik Balcioğlu and Gülsüm Baydar (eds.), *Nesneyi Okumak/Deciphering the Object* (Izmir: Yaşar Üniversitesi, 2012), 36–46, here 37.
55. Simon Frith, *Sound Effects: Youth, Leisure, and the Politics of Rock 'n' Roll* (New York: Pantheon, 1981), 190.
56. Dirk Hoerder, *Cultures in Contact: World Migration in the Second Millennium* (Durham, NC: Duke University Press, 2002), 508–9.
57. Marie-Dominique Lelièvre, *Sagan à toute allure* (n.p.: Denoël, 2008), 121–22.
58. See for the following Mom, *Globalizing Automobilism*, Ch. 3.
59. Khairullina and Santos y Ganges, "Tram, Trolleybus and Bus Services in Eastern-European Socialist Urban Planning," 34.
60. Lyubomir Pozharliev, *The Road to Socialism: Transport Infrastructure in Socialist Bulgaria and Yugoslavia (1945–1989)* (Göttingen: V&R Unipress, 2023), 126.
61. Rostow, *The Stages of Economic Growth*, 87.
62. Björn Wittrock, "Modernity: One, None, or Many? European Origins and Modernity as a Global Condition," *Daedalus* 129, no. 1 (Winter 2000): 31–60, here 55.
63. Wilfred Owen, *Strategy for Mobility* (Washington, DC: Brookings Institution, 1964), 98.
64. Daniel Lerner, *The Passing of Traditional Society: Modernizing the Middle East; With an Introduction by David Riesman* (New York/London: Free Press/Collier-Macmillan, 1965), 49.
65. Michele Alacevich, *Visualizing Uncertainties, or How Albert Hirschman and the World Bank Disagreed on Project Appraisal and Development Approaches* (n.p. [New York]: World Bank, November 2012), 51.
66. See Mom, *Pacific Automobilism*, 159–60.
67. See on this development especially Arturo Escobar, *Encountering Development: The Making and Unmaking of the Third World* (Princeton, NJ: Princeton University Press, 2012).
68. José Ignacio Vielma-Cabruja, "Transfers, Influences and Experiences: Notes on the Development of Expressways in Caracas (1947–1973) and Its Actual Expression" (paper presented at the workshop "Transport and Mobility Studies in Latin America," School of Architecture and Urban Studies of the University of Chile, Santiago de Chile, 8–9 July 2013).
69. Hoerder, *Cultures in Contact*, 499–500.
70. Robert Heinze, "Fighting over Urban Space: Matatu Infrastructure and Bus Stations in Nairobi, 1960–2000," *Africa Today* 65, no. 2 (Winter 2018): 2–21, here 7.
71. Daniel Frederick Vukovich, "Sinological-Orientalism: The Production of the West's Post-Mao China," dissertation (Champaign: University of Illinois at Urbana-Champaign, 2005), 1 (my italics), 59 (the last quote is from Ralph Litzinger [2001]).
72. Ravi Arvind Palat, "Beyond Orientalism: Decolonizing Asian Studies," *Development and Society* 29, no. 2 (December 2000): 105–35, here 119.
73. Heinze, "Fighting over Urban Space," 8–9.
74. Peter R. Gould, *The Development of the Transportation Pattern in Ghana* (Evanston, IL: Northwestern University, April 1960), 135–39.

75. Gina Porter, "Reflections on a Century of Road Transport Developments in West Africa and Their (Gendered) Impacts on the Rural Poor," *EchoGéo* 20 (April–June 2012): 1–14.
76. Adrian Peace, "The Politics of Transporting," *Africa* 58, no. 1 (1988): 14–28.
77. See for the following Joshua Grace, *African Motors: Technology, Gender, and the History of Development* (Durham, NC: Duke University Press, 2021), Chapters 2 and 3 (82–184).
78. Ibid., 154.
79. Joshua Grace, "Heroes of the Road: Race, Gender and the Politics of Mobility in Twentieth Century Tanzania," *Africa* 83, no. 3 (2013): 403–25, here 420.
80. David Wightman, *Towards Economic Cooperation in Asia: The United Nations Economic Commission for Asia and the Far East* (New Haven, CT: Yale University Press, 1963), 204.

CHAPTER 4

The Global Shift to Pacific Automobilism and the Epidemiology of the Car, 1970s–2010s

This chapter covers the half century since the early 1970s, investigating whether the two 'oil shocks' of 1973 and 1979 (suddenly increasing oil prices, first fourfold and then threefold) formed 'turning points' in the history of world automobilism.[1] A first superficial answer would be negative: between 1970 and 1990 world energy consumption *per capita* remained more or less stable (soon after the second crisis consumption values had regained pre-crisis levels), but the doubling of consumption in the developing world (to the level of the OECD countries) suggested that beneath the surface of 'clean' averages, a worldwide shift was occurring.

One of these shifts took place in mobility. After European countries had seen their car densities triple during the 1960s, they increased by another 80 percent during the following decade (partly due to an increase of the single-person household), reaching half of the density values of the US by 1980. A decade later the size of the car fleets of the US and Europe had become more or less equal (at 159 and 156 million vehicles respectively). Automotive 'cores' and 'peripheries' started to shift as well, their borders blurring: Spanish densities between 1952 and 1974 increased fortyfold, to values half of those of the UK, and urban densities in São Paulo were about equal to those of New York. By then, one-fifth of the world's energy was consumed by transport, but such averages hide large disparities: a quarter century later, the share of transport in general energy consumption had risen to one-third, but in Greece and Spain this was nearly 50 percent. Passenger cars were (and are) very 'energy intensive' (in joule per passenger-km), much more so than the bus and the train, and only slightly less than the plane (because of the latter's much higher occupancy).

Shifts also occurred in the modal configuration: two surveys in the UK in 1972/73 and 1975/76 revealed that if *all* person-trips were included (also those on foot, averaging about 1.3 km), walking represented nearly half, and the car somewhat more than one-third. But by 1986 car trips dominated with 51 percent (without walking this increased to 61 percent). It is in this period of what critical transport experts Peter Freund and George Martin have called 'hyperautomobility' that commuting by car became a dominant phenomenon in the West.[2]

Popular histories of this development tend to believe in 'conspiracy theories,' such as what in historiography has become known as the 'Snell hoax.' In 1974, Bradford Snell, staff attorney of the Senate Judiciary Committee, used the energy panic of the 1970s to accuse General Motors of stealthily killing public transport, but GM's lawyers managed to discredit Snell because he could not substantiate this accusation. Less well known is that in 1947 a group of car manufacturers, among them General Motors, and oil companies *were* convicted for conspiracy because they stealthily, through acquisitions and mergers, had converted electric tramlines to lines of gasoline and diesel buses. Although there is ample historical evidence that in the West tram companies were badly managed and often undercapitalized by local governments (leading, in the Netherlands for instance, to their abandonment, without any 'help' from General Motors, as we saw in Chapter 2), transport experts and historians alike are haunted by this controversy until this very day, all the more so since Robert Zemeckis in 1988 released his movie *Who Framed Roger Rabbit?* (and not *Who Shot Roger Rabbit?* as I wrote in an earlier publication in a Freudian lapse). Recently, Nicholas Bloom called the conspiracy idea "nonsense," because the stealthy conversion included forty-five systems, "a fraction of the 1,200 transit systems nationwide." Likewise, Katherine Goodwin added the argument that New Deal legislation forced electric utilities to break up their monopolies.[3] From a historical point of view it is better to view this development against the background of the so-called 'coordination crisis,' the global struggle of more than a century, between rail and road. The shift of the former to the latter is not the result of a conspiracy but of a half-century-long shift toward a capitalist service economy.

The explosion of automobilism was enabled by deep adjustments in the *car system*, from the enormous expansion of road and freeway network building, via the celebration of the 'freedom of car mobility' in movies and general culture, to adjustments in the car's technology. Whereas later in this chapter we will dedicate much space to the first two elements, here we will briefly illuminate two 'revolutions' in the latter: in car design and electronics. Whereas historian David Gartman observed how during the 1950s in the US "the qualitative differences in engineering, construction, and aesthetics between the automotive classes had virtually disappeared, replaced by small quantitative differences in size, horsepower, chrome, and price," General Motors' top designer Harley Earl, as "a postmodernist before his time," added "oversize curves, non-functional grilles, fake portholes[,] express[ing] a fantasy of affluence, luxury, sensual delight and power," to the car, which, in the case of Cadillac, was put on the market in eighty-one models, among which even the American suburban blue-collar worker could "move up economically but not culturally," according to the brilliant analysis of Shelley Nickles, who followed Victoria de Grazia and Lizabeth Cohen's plea to bring class back into consumption history.[4]

The second 'revolution' (which in fact took more than half a century) through electronic control systems took over more and more functions that before the war had belonged to the proudly guarded domain of the masculine motorist. The 'dethronement' of the 'sovereign captain' of the car into a pilot performing tasks suggested by an ever 'smarter' car (against the price of a diminished 'immersion' in the intricacies of handling the moving car) enabled other, less skilled motorists all over the planet to join the 'movement' of automobilism.[5] This process started well before 'electronification' and can be seen as a power struggle between the engineer (as a representative of the manufacturer) and the knowledgeable user, who saw the car under her hands evolve into a foolproof (*narrensicher*) contraption, a derogatory qualification still reminiscent of the eventual victory of the engineer in this struggle, a victory that was sealed by a gradual shift from the 'software' (of the human body) to the hardware (of the car). Electronification (also in the realm of production, allowing the development of sophisticated, fine-tuned suspension systems, engines, and dashboard consoles, to name only a few) supported the shift toward the car interior as an 'entertainment center,' for children during long holiday trips but also for navigating an unknown city or cruising on the freeway with one's feet from the pedals. In other words, the 'loss of the sublime' of car travel has a technical base: it enabled the shift from a vertical to a horizontal transcendental experience, from feeling godlike to being transferred to a parallel 'reality,' as we will see later.

In general, curves that describe the development of car densities during the 1970s and 1980s all over the world do not show large quantitative disruptions in steadily increasing car possession and even use. Yet the 'oil crises' functioned as a trigger of earlier trends, and as an alibi for governments to implement restrictive measures and, later, to raise energy prices: already before the Yom Kippur War of October 1973 speed limits were announced in France, partly as a response to the warning of a 'planetary emergency' by the Club of Rome from 1968.

Indeed, environmental concerns predated the energy crises, although initially, emerging environmental activism showed a remarkable modesty toward automobilism, in the US as well as in Europe. In every Western country, however, an intellectual avant-garde started to formulate an ecological car critique, from journalists and publicists (such as John Keats, Terence Bendixson, René Dumont) via psychologists (Alexander Mitscherlich) and philosophers (Jacques Ellul) to political activists, some of whom (Ivan Illich and André Gorz among them) were able to transnationalize this critique. Whereas Gorz called the car an 'antisocial luxury,' Illich saw the bicycle as the vehicle of the future, quite contrary to President Johnson's White House aid, which promised American citizens that the country, despite the crisis, would not 'go back to the cold, the dark, and the bicycle.' What the crises thus provoked was a mood change about the car among the opinionated part of the Western middle class, and beyond. In 1970 (well before the energy crises) nearly all Dutch respondents to a poll agreed that 'severe measures' against pollution were necessary. No coincidence, then, that world systems historian Immanuel Wallerstein placed the origins of "unsettle[[ing) the structures of the world-system" in the "world revolution of 1968" and Middle Eastern Studies professor Timothy Mitchell speaks of "the 1967–74 crisis."[6] The 'oil shock' added to the public unease: in the US especially, feelings of 'America's decline' spread among the public, which accused the oil companies and the car manufacturers of making common cause with the government in organizing the oil depletion.

The richest regions in the West were the most critical. Already in 1960, California installed a Motor Vehicle Pollution Control Board (seven years later replaced by the still existing California Air Resources Board, CARB), and President Johnson had Congress accept a Motor Vehicle Air Pollution Control Act in 1965. The Environmental Protection Agency (EPA), installed by President Nixon in 1974, was able to force the industry to decrease its car pollutants by 90 percent, and managed at the same time to diminish its fuel consumption considerably. In Europe, a technical as well as political competition between proponents of the 'American' catalytic converter (an 'end-of-pipe' solution) and so-called lean-burn strategies (in which the combustion process itself was adjusted) ended in the victory of the former (cemented in the directive 88/76/EEC, more lenient than the American US 83 standards), mostly because of the American export interests of German manufacturers whom the German environmentalist movement did not dare to oppose. In 1973 Japan too introduced regulation, the most stringent in the world, resulting in a strong preference among citizens for very small cars.

The mood change was also visible in the Western resistance movement against ever continuing and even increasing roadbuilding. This too started in the US with protests against the destruction of urban districts (recent estimates suggest that one million people were displaced during the two decades after 1957),[7] but such protests were soon also inspired by ecological arguments. As early as 1966 protesters in San Francisco successfully opposed the construction of the Embarcadero Freeway, followed by protests and riots in Milwaukee (stopping a lakefront freeway), Baltimore (preventing Interstate 70 from crossing Leakin Park), Nashville (Interstate 40), and Miami, but when President Nixon included one billion dollars in his Federal-Aid Highway Act of 1973 to support public transport, nineteen billion dollars were reserved for the completion of the Interstate system, which at that moment was 98 procent ready, at a cost of one hundred billion dollars. Historian Brian Ladd has argued that the freeway revolts did not 'grow directly out of earlier animosity to cars,' but a critical contemporary of the protests found them inspired by the middle classes' dislike of sharing the road with the workers.[8]

In Europe, too, anti-road protests radicalized when the national systems were nearly ready and national governments opted for so-called 'smart growth,' an electronically driven more efficient use of the existing infrastructure. Radicalization was most outspoken in the UK where the 'civil war over roadbuilding' culminated in the media spectacle of the Battle of Twyford Down in 1992, accompanied by acts of 'eco-sabotage.' Elsewhere in Northwestern Europe, protests focused on the defense of urban integrity against the 'technocracy' of the road engineers, such as in Sweden (where a Traffic Group occupied a parking lot) and the Netherlands (where hippy anarchists in Amsterdam introduced a free-to-use White Bicycle Plan that gained some international fame). Whereas in the US Jane Jacobs's *Death and Life of Great American Cities* (1962) formed a source of inspiration, in Europe transport expert Colin D. Buchanan's report *Traffic in Towns* (1963) for the British government provoked, translated in nine languages, a transnational debate about the 'urban crisis.' During a conference in the Netherlands dedicated to the report, one critical urban planner held a plea in favor of a 'pedestrianization' of Dutch cities, after he had observed that European cities were 'less devastated by the war than many American cities by motorization.'

A third trend (next to environmental and anti-road mood shifts) predating the energy crises was the *road safety* issue: road fatalities during the 1950s and 1960s had increased so precipitously that Western governments started to worry, against a Cold War background, about the legitimation of a form of capitalism in which private ownership of cars was quintessential.[9] It was American President Richard Nixon who in 1969 invited the military association NATO to supervise an Experimental Safety Vehicle (ESV) project to develop a super-safe car. The project failed, not only because the proposed 'technical fixes' made the ESV prototypes much too heavy, but also because a simultaneously emerging international cohort of safety experts identified adjusting drivers' behavior rather than car technology as the most effective solution for a decreasing trend in fatalities, a downward trend that indeed started to occur in most Western industrialized countries from the beginning of the 1970s until well into the twenty-first century. By then, the development toward 'Zero Vision,' the promise of no fatal road accidents at all, tended to level off or even stagnate.

All this is confirmed by Robert Lifset's conclusion that the 'oil crisis' (which hid two other crises: a 'gas crisis' and a 'utility crisis') would have occurred in the US "*with or without* the [OPEC] embargo."[10]

During the 1970s, these environmental and safety concerns seemed to put a stop to the 'exuberant' car culture of the previous period. Automotive historian James Flink observed a turnaround in the American mood (a 'disenchantment') about the 'arrogant' Big Three's (General Motors, Ford, Chrysler) resistance against national antipollution and safety legislation, a mood shift expressed in the popularity among Western middle classes of European and Japanese 'compact cars.' And although the gas-guzzling American car became half a ton lighter and 50 percent more fuel efficient between 1975 and 1995, it never reached European values, because the industry found a loophole in the legislation allowing the development of trucks cloaked as passenger cars, the so-called Sport Utility Vehicles (SUVs), a quintessentially neoliberal car that—quite perversely—promised to guarantee safety to the detriment of lighter 'collision partners.' The history of the automotive middle finger to mobility decency still has to be written, but its military origins are undeniable, starting with the American Jeep and continued by its postwar successors Hummer and Humvee (the latter bought by twenty armies in 2000).[11]

The mood shift of the 1970s can also be observed in a proliferation of the diesel engine in Europe (rejected in the US because of the cheap gasoline there, as well as health arguments and the image of its lower power) and a revival of electric vehicle enthusiasm in both Europe and the US.[12] In fact, this was a part of a more general search for alternative propulsion systems such as gas turbines and Stirling engines. The activities were most hectic in Southern California, the largest car market on earth, haunted by a decline of the military-industrial complex, which set myriad engineers free to develop alternative solutions, stimulated by pressure groups such as the Electric Auto Association and the amendments of the federal Clean Air Act of 1967 and the founding of the Department of Energy. The program was canceled by President Ronald Reagan in 1981, but a decade later new amendments of the Act by President George H. W. Bush enabled the Californian mandate to emerge, to enforce the development of Zero Emission Vehicles (ZEVs), creating a worldwide hope among proponents of an electric future. This in its turn brought a new actor onto the stage: the social sciences.

In Europe national fleet trials were organized, on the island of Rügen in Germany, in Mendrisio in Switzerland, and in La Rochelle, France, with small 'city cars' built by large and small manufacturers. The latter was the result of a difficult start some decades earlier, when the national electric utility company EDF managed to capture existing scientific research on fuel cells and redirected this to research on electric vehicles, but it failed to engage the French state for the project. The analysis of this failure marked the beginning of a new branch of sociological research into scientific and technological innovation called actor network theory (ANT), in this case undertaken by Michel Callon.[13]

⟦ Associated mostly with sociologists like Bruno Latour, John Law, and Michel Callon, and tributary to the SCOT approach (see Chapter 1), actor network theory was developed to uncover and trace "the many connections and relations among a variety of actors (human, non-human, material, discursive) to allow particular actors to become what they are" through a process of "heterogeneous engineering."[14] Aimed at "elid[ing] the distinction between . . . structure and agency," actors (or 'actants' in ANT terminology) are network effects and should be considered a "source of action" rather than entities opposed to 'structure.' It is this aspect that caused a lot of controversy among social scientists.[15] The 'agency of things' became famous through Latour's concept of 'delegation' in which artifacts exercise social control, such as the road bump that forces motorists to drive slowly, thus making a police officer superfluous.[16] ⟧

After the energy crises of the 1970s and stimulated by what then was called the 'electronics revolution' as well as research on new battery types, the electric vehicle (EV) got a new chance to escape from its seemingly eternal characterization as a vehicle of tomorrow, a promise for ecologically conscious segments of the public and at the same time an alibi for the industry to not seriously engage in its development beyond the lucrative subsidies by national and local governments. So-called 'life cycle assessments' (LCAs) enabled comparisons of the emission and energy consumption of the alternatives vis-à-vis a standard gasoline car calculated over the entire 'chain' of production, use, and recycling. Such 'cradle-to-grave' calculations (often limited to 'well-to-wheel' calculations, so without the subsequent end-of-life phase, including recycling) were controversial because the built-in uncertainty factors could be manipulated, which made Shell LCAs immediately recognizable because they predicted the diesel car's emission being preferable over the electric.

In California, General Motors' EV1 (more than a thousand of them) was after two years taken back from its lease contracts and scrapped after the industry had convinced the local authorities that EV technology was not yet market-ready. Researchers seeking a way out of the gasoline car paradigm grew disappointed, all the more so because (as we saw in Chapter 1) the electric alternative was market-ready from, say, 1910, if stakeholders had been willing to install the existing charging infrastructures on a countrywide scale.

But the electric passenger car of the turn of this century was not developed with a 'script' enhancing a more planned travel behavior. Instead, the EV1 had set a trend in another, familiar direction as it was consciously developed as a 'sports car,' an electric 'adventure machine' mimicking its gasoline rival, "sporty, aggressive and clean."[17] It was based upon what David Kirsch has called 'the better battery bugaboo,' the tendency, beginning with Thomas Edison and his nickel iron battery in the 1910s, to see the EV as a 'copy' of its adventurous rival.[18] And although the third and fourth generations between the wars and the fifth generation shortly thereafter were developed as in-factory forklifts and minitrucks, street delivery vans and milk floats (the latter two outperforming their rivals in urban stop-and-go work), the Tesla, considered by many to be at the base of an impending 'EV revolution,' outperformed the EV1 by the adoption of a lithium ion battery (developed by the Japanese company Sony in the early 1990s for an application in consumer electronics), with a much higher energy density, enabling extending the range on one charge considerably. After more than a century, the better battery bugaboo had been made to work.

Based on a neoliberal trickle-down argumentation, the Tesla was offered to an automotive elite who often could profit from national subsidies on their purchase. It was developed from scratch in an amazingly short time by a team of engineers hired from the existing car manufacturers, supplemented with electrical and IT engineers from Silicon Valley firms: the first Tesla Roadster had taken three to five years to bring onto the market by 2006, with an investment of an estimated one to two billion dollars by Silicon Valley entrepreneur Elon Musk. This not only meant a *démasquée* of the industry (equipped with much smaller R&D

departments) with its long gestation times of new cars, it also meant that other tech companies could try the same as well, opening a vista onto a possible future revolution in the automotive industry from electro-mechanical to digital-electronic specialists, "from [a] product-based to service-based business model."[19] This perspective became more real when Musk managed in 2009 to acquire a loan of nearly half a billion dollars from the American Department of Energy, when his company balanced on the brink of bankruptcy. The Tesla Model S (introduced in 2008, an allusion to Henry Ford's Model T; like Ford Tesla's production employed a very high level of vertical integration) not only enabled a range of nearly 500 km on one charge, the simultaneously developed fast-charging network rivaled the combustion car's long-range deployment, while online software upgrades enhanced the car's technical parameters. Tax breaks for Californian buyers supported Tesla's breakthrough further, inciting Daimler (Mercedes) to buy a 9 percent share in the company (followed by a 5 percent stake by China's Tencent). In fact, Chinese tech companies such as BYD, Qiantu, and NIO had already chosen to follow a similar path of luxury EV sports cars, hoping to enable their country to 'leapfrog' Western traditional car expertise. In 2020 Tesla held the highest market share of EVs worldwide.[20]

It took another generation before the electric passenger car would be forced upon a still unwilling motoring constituency, using a deep energy and security crisis in Europe in 2022 (see Chapter 5), although some observers position the tipping point, quite optimistically, in 2005, with the marketing success of the hybrid Toyota Prius (one million sold during the decade since its launch in Japan, by then produced at a hundred thousand per year; fifteen million sold in fifteen years worldwide) and the popularity of EVs and hybrids among fleet operators, from taxi companies to leasing.[21] Before this crisis, Norway had already taken the lead: rich because of its exploitation and exports of massive underwater oil reserves, and eager to start a national industry of electric vehicles (which did not succeed) fed by its hydropower, from 2012 it managed to equip the wealthy part of its middle class with expensive Teslas (the letters of its models forming the word SEXY), followed in 2018 by efforts to market the cheaper (and soon best-selling) Nissan Leaf.[22] In China, sales of EVs really took off from 2014, reaching three million (including hybrids), a production of one million per year bringing the share of these vehicles to 2.2 percent. Marketing research suggested that the predominant buying motive in China is price, while range anxiety does not seem to play a role, let alone the promised lack of local emission.[23]

The mood shift of the 1970s also shines through in mainstream belletristic literature of this period. The protagonist in John Updike's *Rabbit is Rich* (1981, covering the 1970s) expresses a moderate environmental consciousness (often reduced to a notion of energy efficiency), but here too the threat of returning to the bicycle is equaled to the world's end. At the same time automotive adventure felt out of place. But whereas in Joan Didion's postmodern fiction (for instance in *Play It as It Lays*, 1970) such feelings are 'flattened' to a form of freeway 'narcosis,' ecological car critique is nowhere to be found. The same applies to later postmodern American mainstream autopoetic novels (or autopoetic scenes therein) where the car even disappears by a clever technique we have come to call 'absent car' (see Chapter 1, where we called this a rhetorical device to not mention the vehicle although it is always there). In Thomas Pynchon's *Bleeding Edge* (2013), for instance, the 'absent car' is present amid a plethora of urban mobility modes, from walking to taxis and the subway, and the adventure is not so much found in driving but rather in the virtual realm of gaming. Gary Shteyngart's *Super Sad True Love Story* (2010) hardly mentions cars in his depiction of the 'graceful decline of [the] US' and its impoverished middle class members, whose mobility is virtual in the communication through their 'äppärät.' In Alex Shakar's *The Savage Girl* (2001) we also see multimodality (subway, taxis, elevators, rollerblades, and even 'go-peds' and 'peoplemovers') and 'absent cars' present enough to produce 'exhaust,' while the protagonists exhibit a lifestyle of post-irony.

[[The "postironic syndrome," or post-irony for short, was defined by Lukas Hoffmann as literature's "attempt to communicate with the reader instead of presenting her a *passive* entertainment," an attempt he identified to be present in the nonfiction literature of American writers David Foster Wallace and Dave Eggers, representatives of the so-called New Sincerity movement, but others also identified "peripheral realism" in the Global South as an expression of this mood.[24] Writing "straightforward realism," post-ironist writers "are actively struggling with both postmodernism and irony" (the former often called the "age of irony," see Chapter 2), a struggle implicitly directed against 'universal ironists' like Richard Rorty, who can be considered the philosophical spokesperson of this version of literary criticism. Rorty claimed that "the opposite of irony is common sense," and, indeed, 'straightforward realism' is belletristic literature's 'common sense.' But post-ironists propose a counterclaim that the opposite of irony is post-irony.[25]]]

Similar scenes of multiple mobilities (including the use of the bicycle and the bus) are to be found in European 'urban novels' (such as Helene Hegemann's *Axolotl Roadkill*, 2010, or blogger Airen's *Strobo*, 2009, where the young protagonists either don't use cars at all or declare their total lack of interest in them). There, too, post-irony emerged, for instance in Virginie Despentes's *Vernon Subutex One* (2015), fully immersed in a (largely 'absent') car society, in which the protagonists mostly move on foot. The fictional 'absent car' rhymes, so to speak, with the findings of contemporary social scientists, who observed the emergence of the car as an 'ordinary object' (Tim Dant and Peter J. Martin) or the car's 'normalization' and 'near-banalization' (Sven Kesselring).[26]

However, the 'adventure machine' is more easily found in European mainstream novels of this period (for instance in Jochen Schimang's *Das Ende der Berührbarkeit* [The End of Touchability, 1981], or Iain Banks's *Dead Air* [2002]). Still, the modesty in the depiction of fictional mobile adventurousness should be contrasted with psychological surveys (so-called motivation studies) in which the 'lust of fast driving,' the flight experience, feelings of aggression, and other adventurous tropes are invoked by car drivers. Such a 'covert culture' (in the words of Leo Marx)[27] questions the usefulness of mainstream literature as an exclusive source for the motivation of 'real-life' motorists. This is not to deny the appearance of extra-urban car adventurousness in some American novels from this period, such as Steve Tesich's *Karoo* (1998), in which what I would like to call 'body language' (whereby the metal body of the automotive cyborg expresses the emotions of the 'meat body,' such as anger and aggression) is depicted, and the car appears as a 'medium' rather than a transport contraption. In general, however, car driving in Western mainstream autopoetic novels of this period is described either as utterly boring (Richard Ford, *The Lay of the Land*, 2006) or as so ubiquitous that it is not worth mentioning that road trips are 'done' by car (the 'absent car'). There are exceptions, of course, such as the 'road novel' *The Music of Change* (1990) by Paul Auster, in which an old-fashioned 'freedom of mobility' through car possession is preached.

There are also pockets of an adventurous car use in mainstream literatures, however, in what one could call *emancipatory autopoetic novels* written by (or about) women, Blacks and Native Americans, the physically impaired, and even children. Literary scholars such as Heidi Macpherson, Janis Stout, Katherine Mills, and Marie Farr have analyzed women's or other marginals' 'escape narratives' in which classic car adventures are mobilized for emancipatory purposes, introducing relatively new adventurous elements such as 'passengering' and in general more attention to 'life' in the car's interior (Michelle Carter's *On Other Days While Going Home*, 1987; Mona Simpson's *Anywhere But Here* from the same year), making the car into a truly 'ironic vehicle,' because of its cocooning, distancing functions (from the Other). Taxi passengering is celebrated in Karen Duve's *Taxi* (2008, in German) in which all the classic tropes of the 'adventure machine' are mobilized in the depiction of the experiences of the female protagonist during a risky car trip with a journalist in a fast BMW on the autobahn. By then, the passengering phenomenon had also become a subject of study by social scientists.[28]

Whether automotive adventure was transformed when used for emancipatory purposes is open to debate. Pagan Kennedy's *Spinsters* (1995), for instance, fits into the conclusion by Dana Heller (who read more than two hundred novels of women's 'quest romance') that "young women's call to adventure must somehow transcend the limits of an enclosed space: a house, a garden, an institution, an introspective mind," and that the description of "inner turmoil . . . reflects the lack of any outward expression of female aggression."[29] Adventurous women on motorcycles (Rachel Kushner's *The Flamethrowers* [2013], borrowing heavily from her forefathers of motorcycle adventure, including Italian futurist Marinetti [see Chapter 1], or the nationally and internationally very popular Lily Brooks-Dalton's nonfictional *Motorcycles I've Loved* [2015], which made it in 2020 into a film directed by George Clooney) appeared amid a flurry of motorcycle books, still waiting for a specialist in motorcycle culture history to be analyzed and compared to masculine car adventures.

The cocooning ironic car is also to be found in the adventures of fictional Native Americans (David Seals's *The Powwow Highway*, 1979; Louise Erdrich's *Love Medicine*, 1984). Barbara Kingsolver's *Pigs in Heaven* (1993) is a gendered and racialized road novel with a child-passenger as female protagonist. Equally adventurous are depictions of car driving in Harry Crews's (*The Knockout Artist*, 1988) and Larry Brown's 'proletarian' novels *(Dirty Work*, 1989), or the novels dealing with the mobility of 'trailer trash' (Dorothy Allison's *Trash*, 1988, and *Bastard Out of Carolina*, 1992; Sarah Smarsh's *Heartland*, 2018) in which the adventure machine is used as a weapon and car adventure takes on grim, non-symbolic traits. The same applies to the 'migratory subjectivities' in Marilynne Robinson's *Housekeeping* (1980), Maria Golden's *Long Distance Life* (1990), or Toni Morrison's *Song of Solomon* (1977). In Stewart O'Nan's 'emancipatory road novel' *The Speed Queen* (1997) the automotive adventure is even extended into the 'functional' aspects of the world of working-class tinkering and car modification, including the 'feminized machismo' of the automotive beauty contest. In Charles Dickinson's *The Widows' Adventures* (1990) body language is crucial, if not life-saving, when the blind protagonist ('an excellent driver') decides to drive, guided by her sister, to her daughter, living some blocks away. German writer Wolfgang Herrndorf's road novel *tschick* (2010) describes children's learning to drive as a rite of passage full of car adventures, including the affinity between car driving and gaming.

There is not much ecological consciousness in most of these novels, however. This can more easily be found in science fiction novels, analyzed by literary scholars such as Frederick Buell and Heather Houser, in which not only are apocalyptic sentiments revived, but also a rather simplistic car critique can be found based on the conviction that the car has no future and will die out, so why not use it, the protagonists in Edward Abbey's *The Monkey Wrench Gang* (1975) seem to argue, for as long as it goes, for a good purpose? In pulp-fiction writer Philip Wylie's *The End of the Dream* (1972) car production has stopped entirely (as in J. G. Ballard's *Hello America* [1981] after the West's 'oil-depleting catastrophe'), and in German writer Ernest Callenbach's *Ecotopia* (1975), popular among environmental activists of the period, private vehicles no longer exist, and electric buses and cabs are only allowed to drive slowly. Mobility in Marge Piercy's *Woman on the Edge of Time* (1976) is on foot or on bicycles, or by new technologies such as 'floaters' (airships) and air-cushioned 'zoomers' (just like the 'hovercars' in Philip K. Dick's *Do Androids Dream of Electric Sheep?* [1968], made into a movie in 1982 as *Blade Runner*); time travel allows Piercy's protagonist to show her daughter 'a real autocar' that long ago 'killed millions of people!'

If we combine the results of our literary analysis with those of real-world surveys we see a *neoliberal motorist* emerge, a motorist who experiences driving as a matter-of-course, a personal 'right,' its adventurous traits tempered by a modest ecological consciousness, but sometimes mobilized in full for their emancipatory interests. Neoliberalism, authors such as David Harvey and William Davies have argued, emerged in response to the liberation movements of the 1960s (another reason that the 'turning point' has to be placed *before* the energy crises) and found its expression in neoliberal mobility during the postmodern and the subsequent (from the early 1990s) post-postmodern phase, in which "infinitely mobile citizens in search of marketable differences" (in the words of Ginette Verstraete)[30] roam the national freeway networks.

The first four or five postwar decades were not only beneficial for automobilism within the 'First World': the Soviet Union profited so much from the oil price hikes that private ownership of cars per household increased sevenfold between 1970 and 1985, although the car's share in overall mobility performance (in passenger-km) remained low, its average density (the highest to be found in the Baltic states) five to six times lower than those of Czechoslovakia and the GDR. In the fifteen years after 1970, the year marked by historians somewhat optimistically as the start of mass motorization in the "Eastern Bloc," Poland's car density (per capita) increased nearly as fast as that of Hungary (8.5 and 9 times, respectively), closely followed (7 times) by the Soviet Union and

Bulgaria.[31] Exporting one-third of its production in 1972,[32] by 1980 the SU had the same car density as Iran. By then one-tenth of Soviet households owned a car, a share that increased quickly to 15 percent within five years. At the time of the Union's collapse in 1991 there were nine million cars, by far most of them driven by the intelligentsia, most of these Ladas and only 2 percent foreign cars. One-third of the cars were owned by specially selected workers.[33]

Soviet historians have shown that under Soviet leader Leonid Brezhnev, who succeeded Nikita Khrushchev in 1964 and changed his approach of car restraint toward 'mass car production and consumption,' a 'car culture' emerged, but they have so far not found convincing proof of the existence of a 'communist adventure machine.' Probably, however, we will have to find automotive adventure there where we also found it in the Global South: in the proliferation of 'marshrutkas,' privately owned minibuses as alternatives to official public transport, appearing on the streets from the early 1990s, during the perestroika period and driven mostly by immigrants from Azerbaijan and Tajikistan.[34] Until that moment, Mark Keck-Szabel's analysis of Poland as a 'hitchhikers' paradise' suggests that the adventure was more to be found in a special form of passengering, a practice even officially propagated as 'socialist travel,' and also to be found in Rolf Schneider's novel *Die Reise nach Jaroslaw* (The Trip to Jaroslaw).[35]

Only when the Soviet Union and the Eastern European socialist and communist states collapsed around 1990 did mass motorization Western-style occur, most extremely in the GDR, where car densities between 1988 and 1992 nearly (and traffic fatalities more than) doubled, enabled by enormous sums of aid from West Germany. But even then a highly utilitarian car culture prevailed: according to Anna Kruglova, who rejected speaking of a Russian 'system of automobility' à la John Urry, considering the condition of the roads outside the Western urban regions, the corrupt police, and the dilapidated state of many cars, her interlocutors in the city of Perm (one million inhabitants) saw their cars "primarily . . . as a practical possession," and wondered how an owner of a red cabriolet got his spades in the trunk "to dig *potatoes* at the dacha." The postsocialist transition society confirms the crucial role of the middle class, estimated by the end of the 1990s to encompass one-quarter to one-third of the population (its largest contingent consisting of engineers), of which 15 percent were impoverished. The unreliability of Russian statistics often incites social scientists to use car ownership as a characteristic of the middle class. According to this measure, car ownership in 1996 had reached a level of one-third per hundred families.[36] Teachers in St. Petersburg, unable to afford a car after the ruble crisis of

1998, considered a Mercedes "undesirable" because of the 'New Russians' who owned one.[37]

Philipp Ther has analyzed Eastern European neoliberalization as both a conscious construction of a middle class (similar to what shortly thereafter would happen in China, see later in this chapter) and an impoverishment through unemployment and inflation of an 'underclass,' occupying a 400 km wide and 1,000 km long 'zone of poverty' from north to south. The EU extension of 2004 by eight states (Estonia, Latvia, Lithuania, Poland, Slovakia, Czechia, Slovenia, and Hungary) led to unprecedented investments in infrastructure (larger in its share of GDP than the Marshall aid of half a century earlier), which added an East–West orientation to the continent's freeway network.

A Soviet socialist or communist perspective on mobility is more visible in its urban planning tradition, investigated by a new generation of mobility students, one of whom recently made a plea for a correction of the invisibility within mobility studies. Human geographer Martin Müller called this invisible region the Global East, "those countries that occupy an interstitial position between North and South," the former Second World, "too powerful to be periphery, but too weak to be the centre."[38] This new generation of scholars point at a "gap" between planning theory and reality, and have unearthed a similar process of public transport collapse and the highly inconsistent emergence of informal transport (*marshrutkas*, in some cities taking nearly all trips, in other cities nothing) as in the Global South. But they claim, justifiably so, that new approaches are necessary, for instance to focus "on issues related to the production of mobility rather than its consumption," including a "historical perspective," leading to a plea to view subaltern mobilities (such as female street vendors in Georgia's capital Tbilisi) not so much as part of civil society (which would mean "a gross understatement of the difference in political treatment") but as marginal groups with their own political struggles.[39]

When, by the turn of the last century, Japan became the second-largest shareholder of the World Bank (after the US), a neoliberal shift toward the '(Indo-)Pacific' seemed to have been set in motion (expressed, for instance, by the fact that trans-Pacific trade had surpassed trans-Atlantic trade [in value] in 1980), accompanied by speculations about the 'decline' of the US, and even the West as such.[40] Several commentators connect this with efforts by the US government of a flight forward into a kind of automotive and general mobility triumphalism, expressed not only in the YouTube clips showing American soldiers shooting at civilians from the back of their Humvees during their invasion of Iraq under President George W. Bush (2001–9), but also in the seemingly perpetual war of drones by Bush's successor Barack Obama (2009–17). By the end of the last century, a fundamental acceptance of the car society was proposed and promoted by government officials as well as some institutions of social science research, a habitus that Vukan Vuchic has called the "inevitability hypothesis."[41] It is against this background of an acceleration of globalization and neoliberal deregulation ("the relentless reducing of social life to a commodity form")[42] that one can understand the emergence, within circles of social scientists, of the 'mobility turn.'

⟦ The mobility turn was announced in common publications by historian Mimi Sheller and sociologist John Urry around the turn of this century, and was soon joined by geographers, anthropologists, and other social scientists, who declared the need to establish a 'new mobilities paradigm.'[43] Its roots lay outside of transport and mobility expertise, but were based on concepts such as 'liquid modernity' formulated by Zygmunt Bauman, meant to shake up the 'sedentarism' of mainstream social sciences.[44] The result was a plethora of new approaches to transport and mobility under the banner of Urry's concept of 'automobility' ("the unique sociotechnical assemblage that includes cars, road networks, petroleum supplies, and the cultural practices surrounding them," as paraphrased by Rae and Kimberly Quilantang),[45] published, among others, in the journal *Mobilities*, in which, however, historical approaches were rare. Lately, Mimi Sheller added her own (historical) work on postcolonialism as one of the roots of the 'paradigm,' which enabled her not only (nearly a century after Ilya Ehrenburg's similar critique; see Chapter 2) to point our attention again at the devastating extractive aspects of 'the system of automobility,' but also to expand the reach of the movement toward 'mobility justice' and sustainable activism in general, implicitly countering critique that "the experiences of middle class people are treated as normative."[46] ⟧

But automotive history is far from linear. At the same time, in other pockets of the ever more fragmenting Western automotive landscape, we see subcultures emerge that are actively reviving car adventurousness in extreme, over-the-top forms that we may characterize as *carnivalesque* in the sense of Mikhail Bakhtin (see Chapter 3). From the perspective of neoliberalism and its nurturing of societal inequalities we may interpret such revivals as a response to the 'invisibilization' of the working class, whereas the middle class had become "a luxury capitalism can no longer afford," in Slavoj Žižek's words.[47] Examples of such responses are multiple. With a working-class background, the young Swedish *Raggare* or the Japanese Bōsōzoku, the Chicano, Asian American, and Mexican American lowriders, street racers in California, and the car modifiers of the British Midlands, as well as the Scottish girl racers—they all revived two seemingly opposite strands of the original car adventure from the early twentieth century: the 'feminine' and proletarian beautification tradition initiated in the gymkhanas and flower parades on the one hand, and the high-speed thrills of car racing on the other. Whereas the latter brought car adventure to a new neoliberal level, the former celebrated the multimedia and overwhelmingly masculine character of the car through the 'cruising' around of their 'body graphics' (the 'murals' spray-painted on the car body, an extreme form of 'body language') and its spread into the culture of hip-hop videos and TV shows (*Pimp My Ride*, on MTV between 2004 and 2007, interested more in finish than performance). These subcultures (one of the best ethnographies is written by Dag Balkmar in his dissertation from 2012 about boy racers in Sweden)[48] were surprisingly large, extending into the do-it-yourself culture (estimates around 1990 suggest encompassing 10 percent of all motorists) and hot-rod cultures of middle-class (including gay) men and women, whose culture of 'chick cars' is analyzed by Chris Lezotte.[49] Such subcultures, including magazines such as *Glamour* and *Vogue* that started 'automotive advice columns' for women in the 1970s and 1980s or the gay motorists of the Lambda Car Club, founded in 1982 in Atlantic City, are hardly represented in highbrow literature, but rather in so-called pop-cultural examples, such as Erika Lopez's *Flaming Iguanas* (1997, about a female motorcycle ride through the US) or 'popular romance novels' à la former romance novel writer Janet Evanovich (such as *Motor Mouth*, 2006, depicting an extremely adventurous racing culture). Lee Child's *Never Go Back* (2013) about the perfect adventurous car consumer (without a driver's license) Jack Reacher (made into a movie with Tom Cruise in 2016), Viviane Moore's *Le passager silencieux* (2005, The Silent Passenger) about a long-distance street race through Europe, and Johannes W. Betz's *bundesautobahn* (2004, Federal Freeway), a mixture of high-speed adventure and status sensitivity, belong to this genre. They are an active part of a *popular culture* more readily expressed in songs, from pop and rock to hip-hop. Here one can find ecological critique (Joni Mitchell's "Big Yellow Taxi" [1970], David Bowie's album *Lodger* [1979]), next to outright celebrations of car adventure (Bruce Springsteen's working-class reminiscences of watching a 'demolition derby'). Compared to the 1960s (when fifteen hundred 'car songs' were counted in the US), the harvest of such songs in the 1970s and 1980s was modest, until the car-rap culture emerged of Diddy's "Bugatti Boyz" (2010) and gangsta rapper 50 Cent showing off his 'chromes,' a culture in which the status character of (very expensive) cars is celebrated to carnivalesque extremes. The shift to Black performers intensified automotive adventure, just as it did in mainstream literature.

But the true virtual adventure machines can be found in the autopoetic Hollywood movies, their adventurousness analyzed by film students such as Iain Borden, David Laderman, Katie Mills, and Fabian Kröger as masculine emptiness (for instance in Monte Hellman's *Two Lane Blacktop* [1971]). Here, the middle-class scenario writers managed to ironize the car's adventurousness into 'nihilism.' In 'carsploitation films' like Henry Blight Halicki's *Gone in 60 Seconds* (1974) the car chase is celebrated, whereas in between the two oil crises of the 1970s misogynist car movies emanated from the New World production company, founded by Roger Corman, who in the mid-1970s observed that the American public preferred cars after it previously had demanded 'blood and breasts.' Feminist critic Sherry Sonnett Trumbo called such violent movies 'obscene.'[50]

Film studies' extra-urban car adventure bias (culminating in the Australian *Mad Max* trilogy), however, obscures the importance of urban traffic, as depicted in Martin Scorsese's *Taxi Driver* (1976) and in Claude Lelouch's *C'était un rendez-vous* (of the same year), although in the latter case the director's true car vision is exposed in the high-speed Mercedes drive through Paris, with Lelouch's own Ferrari roaring on the soundtrack.

Later, the carnivalesque lowrider and street racing culture became mainstream through the *Fast and Furious* movie series (released from 2001). In many of these popular-cultural utterances the 'ironic car' has disappeared and the adventurousness has acquired blood-serious and sometimes even grim traits. Such virtual adventurousness appeared so powerfully that in other parts of the world rebellious youth tried to re-enact it in the real world, the most extreme example being the Bedouin and Bidun migrant 'drifters' in Saudi Arabia, their extremely status-driven 'high-octane gymnastics' made famous through their appearance in 2003 on the British TV show *Top Gear* and analyzed as a typical, non-ironic plebeian car adventurousness by Arabist Pascal Menoret.[51] Here, as in racing in general, high-speed adventure and status seeking converged into one risky experience.

Another part of popular culture in which the car adventure is celebrated in a carnivalesque manner is gaming, starting with the car racing consoles in the arcades, and culminating in the post-ironic immersion (because of the loss of the distancing effect) in 'dangerously dangerless' and extremely aggressive car games, the most popular being *Grand Theft Auto IV* (2007/8), its development costing $100 million and generating $500 million during the first week after its release. Gaming allows players to create a 'highly adventurous feeling' as an 'out-of-body experience,' according to one of the possible approaches to gaming culture, ludology. Immersion as a new phenomenon (literary scholars now also claim such effects when reading) has been investigated, for instance, through 'close playing' in Jenny Sundén's 'ethnography' of *World of Warcraft* (2004, generating $1 billion per year in revenue, with twelve million subscribers).[52]

One wonders, then, how this fragmented Western car culture was appropriated in the rest of the world, and which 'indigenous' elements were added in the transfer of this culture. One wonders, too, which aspects of this transfer were the result of what Joseph Nye has called 'soft power' and what was hard power; in fact, the entire transfer process can be seen as a form of power between these two extremes.[53] As we have seen in the previous chapter, the transfer process was first and foremost infrastructural, through active road propaganda, enabled by the World Bank, the IMF, and Western consultants. It was, certainly during the first decades, largely a one-way process, known as 'development' (with support from local elites, and sometimes also from other parts of the population). Of the $50 billion spent on transport since its founding in 1947, the World Bank had invested 60–70 percent (the estimates vary somewhat) in roads and highways, but in the 1960s and 1970s this was 80 percent.

By far the most World Bank funding was spent in China, which started to 'open up' in 1978, two years after Mao's death, when Deng Xiaoping gave his famous speech at the Third Plenum of the Eleventh Party Congress. It began with rural decollectivization, the resulting explosion of entrepreneurialism sponsored by the 35 to 50 million diasporic Chinese, dominated by the Taiwanese, and forming the third-largest economy on earth, after the US and Japan. Six years later the start of urban privatization reforms was accompanied by massive migration waves from the countryside. Next to a 'floating population' of more than a hundred million (provoked by illegal, subversive mobility by millions of 'blind flows,' well before the implementation of an urban *hukou* system), a housing reform made other millions of Chinese urbanites relatively prosperous.

Economist Chris Bramall convincingly showed that it was China's domestic, agricultural modernization and rural industrialization that initially drove economic growth, stimulated by the installation of Special Economic Zones in Shandong and Fujian provinces along the east coast, where a gigantic 'workshop of the world' emerged, described for Western readers for the first time by Leslie Chang, who speaks of a migrant population of more than 200 million, many of them highly mobile (if not near-nomadic) 'factory girls,' escaping parental supervision toward a virtually rightless existence.[54]

Industrialization had enormous environmental consequences. As Stevan Harrell has argued in his ecological historical overview of China's 'development,' Marxism inspired a similarly aggressive modernization policy as Western Rostowian ideology, at least until 1998 when several environmental disasters such as floods, dried-up rivers, and industrial waste polluting rivers forced the state to follow a more sustainable approach. The state openly expressed its concerns about the environment in the Ninth Five-Year Plan (1996–2000), introducing in 1999 a 'Clean Automobile Action Plan' and implementing fuel consumption figures between those of the US and of Europe and Japan. China's struggle against pollution is controversial and its analysis highly political (including the American embassy in Beijing publishing air quality figures), caught between extreme observations such as the economist Bramall's (China's pollution is 'routine rather than exceptional') and Judith Shapiro's (the pollution goes back to Maoist 'extreme human interference in the natural world'). Stevan Harrell's recent assessment steers a course between these extremes, praising the effective ('spectacular') reduction of pollution and smog in urban areas, while at the same time observing the 'ecological injustice' of exporting pollution to the less wealthy Chinese hinterland.[55]

But the spread of the 'car society' over the world was much more than propaganda through the construction of national road networks. Investigating this *transfer* process on a global scale is, in the current state of the art of mobility history, too great a challenge. One obvious example is South Korea's US Military Government, which imported propaganda films for "free people [that] praised private car ownership and self-determined mobility."[56] In China, this process had to start much more subtly. There, the construction of a Chinese 'car society' began in earnest with the Sixth Five-Year Plan (1981–85) emphasizing energy and transport (including massive investments in the car industry). Shortly thereafter the state started to invest in road network building (in the Eighth Five-Year Plan [1991–95]) including a freeway network larger than the American Interstate system, which involved the World Bank on a scale never seen before. In addition, during the first decade of the new century, China realized the world's largest high-speed railway network (7,531 km in 2011), which helped enable, together with a high-speed bus network, the enormous rural–urban seasonal migration waves.

When by the end of the 1980s Soviet and Eastern European socialist and communist states started to collapse, and after protests at Tiananmen Square in Beijing were harshly suppressed, Deng secured the support of the local authorities during his Southern Tour (1992) for a gradual intensification of the reforms, but rejecting the neoliberal 'shock therapy' in Europe and Russia. From then on (at a moment when 37,000 passenger cars were counted, most of them used by government officials and their chauffeurs), a surprisingly precipitous economic growth set in, quite different from what happened in Russia. There, the assembly plants owned by foreign manufacturers saw their potential clients prefer secondhand Western cars, until the government raised import tariffs after the 'ruble crash' of 1998, and Russian manufacturers dominated the car market by the end of the century, the breakthrough of mass motorization since 2003 explained by sociologists as a form of 'new utilitarianism' ("efficient time use").[57] Instead, China set up joint ventures with foreign car manufacturers (analyzed in an exemplary way by pioneers such as Eric Harwitt, Eric Thun, and Greg Anderson) whose production of (mainly large) passenger cars and SUVs was hardly affected by the first truly global financial crisis of 1997. It was the start of what one could call a 'Pacific Era.' By then, Asia's Pacific Rim had overtaken the US as the largest car producer in the world, its industry functioning with an overcapacity of more than one-third.[58]

Zooming in now on the establishment of an automotive culture, several observers found an 'emotionally loaded matter-of-factness about cars,' with a clear preference for large cars, fed not only by emulating the usual street scenes of officials' chauffeured limousines with tinted glass, but also and especially to have enough space in the rear to give friends and parents a comfortable ride during weekend outings.[59] Many of the first car buyers stepped up from the bus and the bicycle, the latter's fleet initially growing faster than motorized vehicles to a staggering 180 million in 1995.

In the first years, passenger cars on Chinese streets were not top-of-the-bill, as it seems that initially Chinese negotiators did not manage to force foreign manufacturers to share their state-of-the-art technologies, but when General Motors in 2009 granted its Chinese joint venture partner more than the usual 50 percent ownership share, it also decided to use world-standard technologies in its New Sail and Buick LaCrosse models, perhaps also to avoid China's quick shift toward electric vehicles. In the same year China became the largest car producer in the world (with 10.5 million cars, while the American market had fallen back from 17 to 10 million), the result of a further surge after China's joining the WTO in 2001. By then, the big-car mania had subsided and many Chinese started to buy small cars and midsized sedans.

China's wish to become 'the world's first "Ecological Civilization"' was already visible in its Eleventh Five-Year Plan (2006–10) in which it not only implemented a tax cut on small cars, but also continued to stress the importance of so-called New Energy Vehicles (NEVs), which up to then were produced mostly by indigenous smaller companies, such as BYD (*Biyadi*) and Geely. Chinese citizens appeared as hesitant to go electric as their Western counterparts (despite financial and other benefits), citing range anxiety, safety, and reliability. And although Chinese automobile policy clearly did not manage to 'leapfrog' the internal combustion car, in 2016 the country became the world's largest electric vehicle producer. It also became the country with the largest car-sharing schemes, providing 7 billion rides in 2017, as well as the largest producer of motorcycles (23 million produced in 2015, 3 million more than the annual car production that year, many of them exported). The estimated 200 to 300 million electrically propelled two-wheelers, often charged on the sidewalk by a cable disappearing into the house, represented a large saving of petrol (but an extra demand of coal-fired electricity).

Expert observers explained China's success by pointing at its trial-and-error method of economic experimentation, and its realizing a constantly tense balance between central steering at the state level and the local and regional levels, representatives of the latter forming majorities in central political bodies. As a result of the balance between central steering and local and regional execution, for instance, Shanghai cleverly managed to delay its local road motorization by one generation through the ultra-fast construction of an extensive and efficient subway network. A critic that tried to influence this balance was Deng Yingtao, director of an economic and cultural research center at Beijing's China Academy of Social Sciences (CASS), who suggested in 1991 to discourage car use and instead build a comprehensive system of bicycles, public transport, and taxis. Only more than a decade later, in 2003, did premier Zhu Rongji concede in a speech at Tsinghua University that Chinese leaders had been as surprised by the population's (big) car mania as foreign observers. It was the expanding 'car society,' however, that through its hunger for road space led to the 'hyper-demolition' of China's inner cities, called an "urban holocaust" by American urban planning professor Thomas Campanella.[60] By the second decade of the new century, Chinese global travelers outranked the Americans in tourist spending.[61] By then, China's export of its development-through-infrastructure building model took on a new shape with the announcement, in 2013, of its One Belt, One Road program (meanwhile renamed Belt and Road Initiative, BRI), involving sixty countries with whom China made agreements to build maritime and land (road and railway) corridors, supported by an Asian Investment Bank, meant to become a rival to the World Bank, with an estimated investment of between one and eight trillion dollars, the "largest overseas investment drive ever launched by a single country."

By then, the 'Asian Highway' project (see Chapter 3) had expanded to more than 100,000 km, evolving into a network of which the largest shares were to be found in China, India, and Kazakhstan.[62] Thus, China became a part of the "global growth coalition" that emerged after the credit crisis of 2008/9, inciting "global capitalist accumulation dynamics" and consisting of "multilateral development banks, multinational corporations, multilateral governmental institutions, consultancies and some of the most powerful governments in the world."[63] In their geopolitical rivalry, termed by Seth Schindler a "new Cold War," smaller states sense the opportunity to connect to transnational 'flows' by "systematic restructuring" (in fact: centralizing) into "infrastructure states."[64]

Who were those car-buying Chinese and what were their preferences? According to Australian sinologist Luigi Tomba the car-buying middle class was consciously constructed through "cultural engineering":[65] shortly before the turn of the millennium the salaries of selected urban white-collar civil servants and other professionals were raised two- to fourfold, after entrepreneurs (many of them with a rural background) had already started to buy cars, trucks, and tractors. This new class (relatively small, non-uniform, and still in the making by 2012, when a major analysis on its formation was published by Chinese scholars)[66] developed close ties with the state, which several observers use to explain the eagerness to 'jump into the sea' (leave their government and university jobs to engage in entrepreneurialism) after the Tiananmen uprising of 1989. Whereas in the West, and especially in the US, motorization started in the countryside in the small towns of the less populated states (see Chapter 1), in China, apart from an initial purchasing urge by entrepreneurs from the countryside that may have meant the breakthrough of car mobility, the 'car society' initially blossomed in the big cities.

If we look into (English translations of) Chinese literature, the classic 'adventure machine' (even in its tamed, Western form) is largely absent: the adventure is in the purchase of the vehicle and its conspicuous consumption (as for instance in Mohsin Hamid's *How to Get Filthy Rich in Rising Asia* [2013] or Tash Aw's *Five Star Billionaire* from the same year). Yu Hua's monumental *Brothers* (2005–6) and Wen Zhu's *I Love Dollars and Other Stories of China* (2007) emphasize the status character of the cars, described without much irony. And although sinologist Beth Notar discovered and interviewed some isolated 'drifters' and other adventurers (such as illegal street racers) in 2008, even popular TV series (such as *Du Lala* and *Tiny Times*, shown between 2007 and 2016, or *Ode to Joy* [sixty episodes during 2016–17]) focus on the *status hierarchy* enabled by the different makes and models rather than the car's adventurous potentials. In these series a hierarchy of cars and fashionable clothes, depicted as a "crucial and active technique of neoliberal subject-making," accompanied the development from petty bourgeois (*xiao zi*) to "white-collar managerial elite woman."[67] Such images resonate with the importance of conspicuous consumption emphasized time and again by surveys of "luxury consumption."[68] Wang Shuo's postmodern novels (*Playing for Thrills: A Mystery* [1989; translated 1997]; *Don Xu* [1997]) resemble the 'urban novels' in the West in their depiction of a plethora of mobility modes, from taxi, subway, and bus to cycling and walking. Wang's female successor, Wei Hui, is one of the few who dedicated some space in her semi-autobiographical *Shanghai Baby* (1999) to old-fashioned car adventure during her and her Generation X friends' midnight spin on the elevated highways, openly flirting, as an alumnus of Fudan University's Chinese Language and Literature studies, with "those carnival-like poetry happenings of the 1960s in the West."[69] In this Chinese version of global "chick-lit" and its copycat versions with their heterosexual quasi-feminist rhetoric (*Canton Baby*, *Chengdu Baby*, in which the female neoliberal subject is celebrated, according to Eva Chen), the protagonist meets Generation Y women at a party who own their own cars, the same youngsters who (according to marketing studies) still mentioned 'safety' as primary buying motive, followed no longer by space-in-the-back but by 'exterior design.'[70] Such youngsters watched the *Fast and Furious* and the *Transformers* series, and discussed on the Chinese internet platform Douban the relationship between the protagonists approvingly, rather than their carnivalesque car adventures. Fifth- and sixth-generation film directors, famous in the West, show drag-racing street gangs and drifters (Zhang Yang, *Driverless* [2010]; Guan Hu, *Mr. Six* [2015]), while famous rally driver Han Han's protagonists in his *The Continent* (2014) escape from a car chase with the police in a Volkswagen Polo during a journey through China; a similar cross-country SUV trip is shown in Ning Hao's *Breakup Buddies* (2014).

As to the motorization in Asian countries beyond China it is at this moment very hard, if not impossible, to distill a coherent pattern, let alone observe the existence of automotive adventurousness as a motive of purchase and use on the basis of literary and popular-cultural sources alone, given the dearth of sufficiently detailed studies by social scientists. Social science has invested quite some investigative energy into the problem of the 'self' in Asian youth, particularly in the issue of individualism understood as selfishness in Chinese youngsters, or the observation by marketeers of the 'emotionally loaded matter-of-factness about cars,' but the findings are so far inconclusive.[71]

Take Vietnam, like China shielded from the World Bank's neoliberal 'structural adjustment programs' because of its communist background, and like China engaging in its own neoliberal modernization called *đổi mới* (renovation), announced by the Sixth National Congress of the Communist Party in 1986, and like China joining the WTO, in 2005. But different from China, Vietnamese citizens motorized through the motorcycle, reaching a density of 1.3 per household in 2014, when the country counted 92 million inhabitants, a density only surpassed by Taiwan, but twice that of Thailand. Constantly opposed by the government (which used the safety argument), some young Vietnamese customized (and women repainted) their two-wheelers, organizing illegal street races. But on the weekends village roads saw the Toyota and Nissan cars of the 'new rich' from the big cities appear driving to their villas hidden in the mountainous areas. When other layers of the middle class also started to buy cars, they did not abandon their two-wheelers, and indulged in a revival of the classic car adventure described meticulously by sociologist Arve Hansen's participative observation (on the motorcycle!) during his fieldwork in 2012 and 2013: whereas the motorcycle was used for utilitarian purposes (such as commuting), cars were used once a week on average, to show off and indulge in 'dangerous high-speed driving.' By then, the World Bank called Vietnam a 'development success story,' at a moment when owning a car had lost much of its status character: "Everyone has one now," a businessman in Hanoi told Hansen.[72]

Or take, by contrast, South Korea, as an example of a so-called 'development state' and one of the 'Asian Tigers' or 'Four Little Dragons' (together with Taiwan, Hong Kong, and Singapore a source of inspiration for the Chinese 'opening up' process; they all motorized faster than China). South Korea increased car production tenfold to nearly one million during the 1980s, exporting its products (dominated by small cars) to more than 170 countries. Its neoliberalization was guided by the same team from Stanford University that had also worked in Latin America's Southern Cone. By the end of the 1990s even workers took part in the car culture's conspicuous consumption led by South Korea's new middle class. According to researchers such as historical anthropologist Denise Lett and sociologist Seung-Kuk Kim, the new motorists were extremely status-sensitive.[73]

In a similar process, the South Korean state set up a world-class IT sector, making the country the core of a worldwide gaming culture. Together with Japan, South Korea exported the cybercafé craze to the rest of the world, including China, where from the early 1980s a virtual adventurous culture was created, performed through half of the world's online games, *World of Warcraft* counting more subscribers in China than in the US.

South Korea was the major beneficiary of aid from another Asian country that grew into a development hub itself: Japan. During the 1970s the Japan International Cooperation Agency (JICA) focused on technology transfer, funding surveys, planning, and construction of roads, railroads, and bridges in Burma, Indonesia, South Korea, the Philippines, Thailand, and Vietnam, but this could not prevent participants in anti-Japanese riots accusing Japan of using aid to secure food and natural resources. In Thailand hundreds of millions of yen were invested in its road network (its car market dominated by Japanese brands).

Japan's 'neoliberal turn' occurred in the mid-1980s when the Liberal Democratic Party started to invest heavily in roadbuilding, after previous progressive governments had been more modest in this respect and instead had waged a 'traffic war' against the alarming road unsafety caused, in Tokyo for instance, by the 'kamikaze taxis' with their 'breakneck speeding.' Japan's car culture deserves more research as it is crucial for a robust understanding of Asia's motorization, not only because its car emission legislation in the 1970s became the most stringent in the world, but also because of its restrictive policy based on the proliferation of small and medium-sized cars, which also found a welcome outlet in other countries, feeding the 'Japanization' of Western car markets and beyond. There, Japan's industrial prestige increased during the 1980s and 1990s when it managed to improve its productivity through 'lean' and 'just-in-time' production methods developed within its car factories. Meant for the domestic car market were the K-cars (K standing for *kei-jidōsha*, 'lightweight,' they do not need a difficult-to-get parking license), some of them the subject of a modifying and street racing subculture.

But in the cities a bicycle culture with Japanese characteristics kept developing. In the year that the Japanese economic bubble burst (1991) 71 million bikes were counted (double the amount of cars), most of them a technical convergence of the *mama-chari* and the mini-bike (see Chapter 3), in a country in which only 3 percent of the 'space' accessible to bicycles was for exclusive bike use only. Still, 15 percent of daily outings are done on the bike, ranking third in the world (after the Netherlands, with 30 percent). Japanese parliamentarians declared at the Johannesburg Summit on Sustainable Development that "The bicycle will save our planet." In the 1990s China became the world's largest bicycle manufacturer (producing four-fifths of the annual 130 million bicycles).[74] Two decades later China's sales of electric two-wheelers amounted to 30 million, annually adding to a domestic fleet of more than 220 million vehicles.[75]

A case on its own is India's neoliberal motorization, started in the early 1990s by Prime Minster Narasimba Rao and his finance minister (and later prime minister) Manmohan Singh, with heavy Japanese investments after decades of prioritizing railway expansion. Although until at least 2010 the motorcycle dominated (when there were ten times more two- and three-wheelers than cars), and a freeway network reached only a total length of 500 km in 2004/5, the emergence of the rural middle class (a product of the Green Revolution) and especially the new professionals in the cities created an exorbitant consumption culture in which the car was only one of the many luxuries, a culture described by several observers as a 'collective neurosis' and a parvenue "opulence," according to Noam Chomsky never seen "even in America."[76] By 2010 eight million cars were produced per year, adding to the local fleet of 122 million motor vehicles, a modest 22 per 1,000 population.[77]

In virtual reality, India's motorcycle did not become an 'adventure machine' à la Hollywood's *The Wild One* or *The Leather Boys*; instead, Bollywood made the motorcycle into an 'instrument of romance,' as in *Bobby* (1974) and *Ek Duje Ke Liye* (1979, Made for Each Other), testifying again to the fuzzy character of an artifact's 'script' (see Chapter 1). In cruel reality, however, the intimidating motorcycle and scooter processions of the right-wing and ultranationalist Shiv Sena urban toughs during India's bloody riots in Mumbai in 1992/93 tell a different story.

India's (as well as neighboring Bangladesh's) mobility is extremely layered, its bottom range formed by the tens of thousands of hand-pulled (in Kolkata and neighboring Bangladeshi Dhaka), cycle, and three-wheeled auto-rickshaws, 40 percent of the latter used by women and often running on LPG or LNG (liquid petroleum and liquid natural gas types), and lately sporadically also equipped with electric propulsion systems. In Delhi (where only 5 to 20 percent of the total vehicle fleet are cars) estimates of the cycle rickshaw fleet amount to 600,000, driven often by inhabitants of the slums. They function as an "interchange" between other modes and for shopping, irrespective of the patrons' class.[78] In smaller Indian cities too road traffic is extremely layered: in Dehradun, at the foot of the Himalayas, for instance, a fleet of 1,600 Vikrams (locally produced oversized auto-rickshaws for up to six passengers) wages a "war" with the equally informal private city bus fleet.[79]

In Dhaka alone 500,000 rickshaws were counted in 2015, all of them driven by men, only one-fifth of them licensed; estimates of the total fleet in Bangladesh's capital go as high as 1.1 million, providing direct work for 800,000 people. Their movement in traffic does not derive from a concept of 'flow' (as cars have been made to move by transport engineers in Western traffic) but from a very rich 'body language':

they "bump awkwardly along creating pathways as they go. They grate against each other, dodge, stop and go, negotiate obstacles, back-track and move off in new directions propelled by different intersecting logics," as geographer Caroline Knowles observed. Such subaltern mobility was banned by Dhaka's police in 2004/5 from certain traffic corridors, on the basis of transport reform funded by the World Bank (which later had to concede that banning such vehicles did not decrease congestion because of the explosion of buses and minibuses). Similar bans were attempted by the communist chief minister of India's West Bengal, Buddhadeb Bhattacharjee (2000–11), to no avail, although streetcorner workers have to remain vigilant in order to preserve one of the few livelihood options of the rural and urban poor.[80] In Dhaka an impressive electrification of cycle rickshaws started around 2007 on the basis of recycled lead-gel batteries, a development resulting in a fleet of between one and four million vehicles ten years later, only to be confronted with a nationwide ban in 2021.[81] Similar bans have been observed in Chinese cities against electric two-wheelers, using safety and flow arguments.[82]

Recently, excellent local studies by historians and social scientists have extended our knowledge of India's (car) mobility considerably, through analyses of the history of 'congestion' in Indian cities, the struggle over street space in Kolkata between road and roadside users (motorists versus street vendors, another example of India's extreme urban mobility layeredness), and an anthropological study of Muslim taxi drivers in Mumbai.[83] In the latter case, Tarini Bedi explained how a family-based culture managed to keep the old Ambassadors running, showing how 'the shock of the old' (David Edgerton) works on the ground through the constant 'modernizing' of this iconic Indian 'people's car.'

Production of the Ambassador was halted in 2014, but meanwhile other indigenous cars had been put on the Indian streets, such as the Maruti (a Suzuki clone, from 1982) and the Tata Nano (2008). India's neoliberal turn enabled the country to become a cog in the global car production machine, when many multinational car manufacturers set up plants in the country and India became a 'production base for global markets of compact cars' with a market share (in number of cars) in 2017 of 5 percent. Malaysia also developed its own car-for-the-common-man: the Proton Saga (a technologically questionable Mitsubishi derivative), but many Indian and Malay middle-class car buyers preferred 'real,' imported cars, and if they chose in favor of the Tata or the Proton, they often opted for less 'basic' models, which were soon added to the product range, a phenomenon one can observe also in the West.[84]

Indian literati reflect the status-sensitive behavior of their compatriots. From Pankaj Mishra's *Butter Chicken in Ludhiana* (1995) with its somewhat condescending description of small-town middle-class mediocrity, to Aravind Adiga's *Last Man in Tower* (2011), the commodity character of the car (as a commuter vehicle, replacing the formerly used auto-rickshaw, train, and shared taxi) seems to be more important than its adventurous potential, but there are also novels that thematize the emancipatory function of passengering (Ratika Kapur's *The Private Life of Mrs. Sharma*, 2015) or add a realistic description of Delhi's motorcycle mobility, including its unsafety (Uday Prakash, *Paul Gomra ka Scooter* [Paul Gomra's Scooter], 2002). India's absence of motorized mobility among the rural lower middle classes can be witnessed in Neel Mukherjee's *The Lives of Others* (2014) or among the revolutionary Naxalites in the North (who bury captured cars for later use) in Arundhati Roy's *Walking with the Comrades* (2011).

Elsewhere in Asia similar literary developments can be observed (see for instance Andrea Hirata's *The Rainbow Troops* [2005] or Eka Kurniawan's *Beauty is a Wound* [2002] about a subaltern mobility culture on foot, horse-drawn carts, and the bicycle in Indonesia, invaded by utilitarian-deployed motorcycles and cars), but perhaps the best way to investigate the possible adventurousness of Asian automobilism is through the so-called 'diasporic novels,' in which 'transcultural subjectivities' are to be found constructing their 'hybridized identities.' The true transnational vehicle in this context is of course the plane, while the car's reach is continental at most. From the Chinese diaspora in Malaya (Catherine Lim's *The Teardrop Story Woman*, 1998) or Canada (Lydia Kwa's *This Place Called Absence*, 2000) to the younger generation of 'digital' and other world nomads (Hwee Hwee Tan, *Foreign Bodies* [1997] and *Mammon Inc.* [2001]) or the Indian, Bangladeshi, African, Caribbean, and Vietnamese diasporas in Great Britain, the US, and Australia (Vanessa Walters, *Rude Girls* [1996], Zadie Smith, *White Teeth* [2000] and *On Beauty* [2005], Monica Ali, *Brick Lane* [2003], Hsu-Ming Teo, *Behind the Moon* [2005], Taiye Selasi, *Ghana Must Go* [2013], Chimamanda Ngozi Adichie, *Americanah* [2014], Chigozie Obioma, *The Fishermen* [2015]), in all these novels the car is used as a tool to create a contrast between 'home' and 'host.' Through the car the 'bipolarity' of global automobilism is investigated, as a token of affluence (also in the West), whereas car ownership and status are emphasized over adventure. In some cases the mobility experience in the 'home' country seems to be more modern than in the diaspora's poor members' living conditions in the West.[85]

Summarizing, motorization in Asia during the half century after 1970 (where in most countries a road system was in place by the end of the 1980s) is extremely varied, but it is quite clear that the middle class (in several cases consciously created by the state as the start of modernization and neoliberalization projects, including the hope of societal 'stabilization' and even democratization) is crucial in the proliferation of the motorcycle and the car. A counterexample is Indonesia, where the economic boom during the 1970s, after the military coup against President Sukarno (1965), did not make the four million government bureaucrats prosperous enough to enable them to buy cars. The extreme opposite is Singapore, one of the examples of China's leaders, which seems to consist of one big middle class, where the government managed to 'tame' the car through its Area Licensing Scheme (ALS, making the inner city into a restricted zone), to enforce centralization of public bus transport and force 'pirate taxis' into a licensed system.

A second conclusion from our 'Asian case' is the importance of status as purchase and use motive. However, as already said, it seems too early to determine the role of adventurousness in mobility, although it is certainly there, both in Asia's car culture and in its motorcycle culture. Asia's autopoetic literature is largely a middle-class affair, and even if it depicts the mobility of the working class or the underclass, informal transport is largely absent. This 'rhymes' with the car in the West: what is ubiquitous does not deserve mention, or even: what is not middle-class tends to become 'invisible.'

How is this in Africa? The continent was hard-hit by the energy crises, its debt increasing from six (1970) to 86 billion dollars in 1983, caused in no small part by "misplaced highway infrastructure investment," according to critical transport engineering professor John Howe, triggering a large brain drain to the West.[86] During the 1980s forty African countries underwent the IMF's and World Bank's 'structural adjustment program' (SAP) treatment, which caused many of the one million kilometers of sub-Saharan roads to return to the bush, despite the development aid, not only from the West, but also from the Soviet Union and especially China, against a Cold War background. Whereas the World Bank had meanwhile changed its policy to poverty reduction (emphasizing agricultural roads and feeder roads over highways), China maintained its focus on large-scale infrastructure. By 2010 it had surpassed the World Bank as the biggest lender on the planet; it also sent thousands of volunteers giving Mandarin courses overseas, joining the 'adventurers' (engineers, medical staff, farmers). By then one-third of China's oil came from Africa.

Other aid providers (from Norway, the African Development Bank, and the UN Economic Commission for Africa) hoped to contribute to Africa's 'economic renaissance' in the 21st century, taking the initiative for a continent-wide Trans African Highways (TAH) network. African governments founded PIDA, an ambitious Programme for Infrastructure

Development in Africa, aimed at expanding the continent's road system with more than 400,000 km of major, smaller, and rural roads. The new roads also allowed mining organizations, largely funded by China, to transport their resources to the coastal ports, such as Djibouti (part of China's BRI project), connected by Africa's first electric international railway through loans of $4 billion from Chinese banks.

The privatization dictates (Nigerian Afrobeat singer Seun Kuti wrote a protest song titled "IMF [International Motherfucker]") not only led to massive layoffs and high unemployment, but made a small minority very rich. As a result, the continent's automotive fleet increased from 9 million in 1980 to nearly 27 million (in both cases around one-third trucks, buses, and tractors), not a few of them imported secondhand on large roll-on-roll-off ships from Northwestern Europe. These cars increased the layeredness of traffic in the extremely fast-urbanizing African countries so much that congestion became a matter of course from the end of the 1970s. The layeredness was also experienced in the layout of the cities, where road modernization took place in the (old colonial) residential areas and the industrial areas, also after independence, whereas migrants lived in the indigenous residential areas, often not included in any transport planning.

A considerable part of this layeredness were the combis (South Africa), *daladala* (Tanzania), *danfo* (Nigeria), *tro-tros* (Ghana), and the thousands of 'informal' minibuses, motorized and non-motorized two-wheeler taxis, and repurposed trucks, handled often in a carnivalesque manner, and sometimes far beyond: historian Kenda Mutongi described the cranking of a *matatu* engine by holding a petrol-saturated cloth over the carburetor, and 'reckless driving' incited passengers to act 'as though they were [in] an amusement park.'[87] In Kenya, the *matatu* culture became highly politicized when, in the run-up to the 1992 elections, it became a stronghold for a multiparty system against president Arap Moi, but then "criminal cartels and youth gangs" captured the sector, implementing their own "criminal . . . self-regulation." When after the 2002 election the winning opposition candidate announced new "regulations, including speed governors and uniforms for the drivers," police harassment did not stop, and "police death squads" even appeared on the streets of Nairobi.[88] Likewise, the taxi wars in South Africa of the late 1980s, fought with AK-47s and costing fifty lives, were beyond Bakhtin. The same applies to the election politics "constituted as warfare" in Nigeria, mafia practices in which informal transport riders and other personnel were closely involved. And yet, the collapse of many local public transport schemes based on a formal bus network, often the result of deregulation enforced by a neoliberal lending policy by the IMF and World Bank, made social scientists (especially indigenous ones) and others who studied the 'informal' response become gradually milder, such as Martin Oteng-Ababio and Ernest Agyemang on the Okada motorcycle fleets (Okada was the name of a Nigerian airline) in Ghana and neighboring Nigeria, Togo, and Burkina Faso. Such vehicles are called *alalok* in Nigeria, *kabu-kabu* in Niger, *boda-boda* in Uganda and Kenya, *zemidjan* in Benin, *oleyia* in Togo, and *bendskin* in Cameroon. In Nigeria's capital Lagos alone 200,000 operators are counted who provide employment for 500,000 people. Informal transport, indeed, is first and foremost a labor issue (hence the sporadic attacks on World Bank-funded Bus Rapid Transit [BRT] in Lagos), but informal vehicles have a double use profile: they are often also used for domestic purposes, for instance as a vehicle for commuting. In Accra (Ghana), again related to the run-up to elections, the government issued a ban on these vehicles, to no effect.[89]

Indeed, after the collapse of many local public transport schemes from 1989, 'informal transport' captured shares of up to 80 percent (and in some cases even 100 percent), around which a popular culture emerged: musicians composed special music for the 'Generation Matatu,' who, while passengering in Chinese and other cheap new minivans, were accompanied by 'deafening noise' and 'carnival barking.' Street children 'minded' such vehicles, often embellished with slogans such as 'zigzag,' 'if you speed I speed,' and 'always ahead,' considered masculine in contrast to the 'feminine' buses.[90] Informal transport recently underwent a boost, superposed upon the normal growth of the population, because of new waves of migrants and the ever increasing practice of migrant smuggling.[91] The overwhelmingly broad emergence of this phenomenon is not only triggered by economic motives, but also by local, municipal power mobilized against higher echelons of government.[92]

Transport policy researcher Paul Barter coined the term 'bus city' for urban conditions largely dependent on formal and informal (mini) bus transport; he later added the epithet 'motorcycle city' to his efforts to characterize the specificities of the urban Global South. But there is more: planning professor Robert Cervero pointed at this culture's influence on Western urban mobility cultures, with their 'black cabs' in Belfast, the car sharing of commuters in San Francisco, and the 'taxis colectivos' of Santo Domingo.'[93] Let us also not forget that jitney cultures (known around World War I) persisted in the US into the 1970s (for instance on the South Side of Chicago) and even longer (in Pittsburgh, "illegal, but thriving").[94] And although the World Bank slowly started to recognize the reality of this enormous, worldwide subversive and often subaltern sector, the introduction of what one could call 'sustainable neoliberalism' opened a new angle of attack on the sector's inefficiency and (two-stroke) pollution.

Africa's autopoetics confirm the adventurousness of 'informal mobility,' although here, too, the status character of car ownership (for instance in Kenyan Ngũgĩ wa Thiong'o's 'development novel' *Devil on the Cross* [1980]) is much more outspoken than in Western novels. His *Striving for the Wind* (1990) even contains a Rabelaisian scene in which the failure of a police raid is answered with riotous 'delirious laughter.' Sihle Komalo's travelogues also deal with over-the-top adventurism (for instance in *Almost Sleeping My Way to Timbuktu*, 2013), long after former Peace Corps volunteer Peter Chilson returned to Niger in 1992 to experience the 'pedal-to-the-floor insanity' of the 'bush taxis' as opposed to the 'boring' car driving at home. Other observers refer to the transporters of the perishable 'Green Gold' of the *khat* stimulant plant, racing at top speed to their clients in Nairobi and other cities, negotiating skillfully the potholes in the road.[95]

Even Tanzania had to succumb to a neoliberal reform of its socialist public transport. Private operators were banned again after a brief period of tolerance (in 1972–74), but in 1989 pressure from the IMF (which withheld the last tranche of its SAP aid) forced the government to 'liberalize' the market, which by the end of the 1990s consisted fully of privately owned (mostly Japanese) vehicles. This was after the local authorities had observed (in 1992) that nearly all fatal accidents in Dar es Salaam were related to the *daladala*.[96] Joshua Grace rightly argued that before privatization, Tanzania's *daladala* could be considered a truly "populist socialist technology, a people's car," the result of a 'new approach' to the car without the scarcity argument of Soviet scholars.[97] This rare experiment was accompanied by a blossoming literary culture of 'development fiction,' analyzed by historian and Africanist Emily Callaci: pulp fiction published between 1975 and 1985 accompanying the country's mass-literacy campaigns, first mainly written by women, but then taken over by male urban youth whose fictional car chases and shoot-outs, Western-style, constructed "a [fictional] male mobility as a sign of Afro-modernity."[98]

Recently, specialist in French literature Anna-Leena Toivanen analyzed several rather rare novels about undocumented migrants in European peripheries, the "petit migrants" in Achille Mbembe's words, "who end up producing non-elitist, practical forms of cosmopolitanism in clandestine spaces." One such novel is Fabienne Kanor's *Faire l'aventure*, the adventure (in the sense of 'trying one's luck' in getting to the European core) that one of the protagonists, Biram, 'does' being quite different from the carnivalesque diatribes by the African fictional (lower) middle class discussed above. Biram gets, after a visit to a fancy nightclub, in "an expensive German vehicle" to his penthouse, but the real 'adventure,' in the boat over the Mediterranean and on his nomadic foot journeys over no fewer than ten years, is only mentioned in short flashbacks, in a way following the style of travel writing scholarship that also focuses on 'home' and the destination. Toivanen also mentions the 'adventure' of a female protagonist as being a prostitute. Such adventures acquire a grim and life-threatening character, which the novel contrasts with the 'adventures' of the tourists in Tenerife.[99] In general young African migrants characterize their deadly journey to Europe as "faire l'aventure" (adventuring).[100]

At the turn of the twenty-first century, Latin America was, after Australia, the world's most urbanized continent, the prosperity of its urban middle classes increasing after the redemocratization of the 1970s. With this, environmental problems proliferated, from Amazonian deforestation (Brazilian union activist Chico Mendes, murdered in 1988, was its most famous victim) to the pollution in Mexico City (where more citizens died of respiratory problems than through car accidents).

In Brazil, the military dictatorship lasted twenty-one years until 1985, its 'Brazilian Miracle' enabled by a combination of union suppression (especially in the domestic car plants of Volkswagen, Ford, and General Motors) and foreign capital, allowing the middle class to motorize precipitously between 1960 and 1980 (by 1.6 million cars). Although the country was hard-hit by the oil price shock of 1973, a *Proálcool* program of dedicated alcohol-fueled cars (running on fuel distilled from sugar cane) alleviated the dependence on fossil fuel somewhat. Brazil is also the country where in many cities the subversive *clandestinos* (second-hand Volkswagen vans in the mid-1990s, later replaced by Japanese and South Korean minibuses) became integrated into the local transit service. They were so successful that regular bus and train services in São Paulo and Rio de Janeiro plummeted by one-third. The 1.3 million inhabitants of the *favelas*, however, remained dependent on the informal motorcycle taxis (often 125 cc Hondas).

A future world history of mobility owes much to Brazilian urban planner Eduardo Vasconcellos, who not only criticized the mobility injustice created by the close 'symbiosis' of middle class and state with its "far from democratic" decision-making process, but also wondered why, in developing countries in general, 'transportation engineers' were not held responsible and were "not in practice prosecuted for traffic accidents in their jurisdiction," just like structural and industrial engineers if they made serious mistakes.[101] The answer, I venture, is as cynical as it is true: because traffic deaths are not considered mistakes.

Latin American autopoetics from this period (such as Horacio Castellanos Moya's *The Dream of My Return* [2013; translation 2015] and Fernando Vallejo's *Our Lady of the Assassins* [1994, made into a movie in 2000]) added the murderous character of the car-as-weapon (and in the latter case: motorcycle-as-weapon) to the adventurous traits acquired in previous periods, no surprise for anyone who is aware of the deployment of the car in military repression (such as the Ford Falcon in the 1970s, preferred by the Argentinian death squads). Or literati used surrealism as an escape from cruel reality, such as Roberto Bolaño in his monumental *The Savage Detectives* (1998; translation 2007), who not only reinvented true Baudelairean automotive urban *flânerie*, his protagonists' extra-urban car adventures also border on the surreal (a truck 'looked like a burned arm').[102]

The globalization of the 'car system' into a worldwide 'car society' that took place in the period covered by this chapter allows us for the first time to throw a (albeit preliminary) *transnational and global* light on the world's mobility. This promises to provide us with a vista superposed on the fragmented vision of the previous chapters, resulting from the local and regional patterns that could be derived from a concatenation of national stories.

Such a transnational vista may start with constant expansion of a worldwide road system (in an estimate by the CIA reaching 37 million km). Yet the World Bank's investment in infrastructure slowly declined, and dropped suddenly in the mid-1990s, because of an increasing interest from private investors (possibly related to roadbuilding, as this is exactly the time of the expansion of the global road network).[103] Worldwide (that is: fueled first by the West) a doubling of global car production occurred between 1960 and 1975. In 1978 world production (31 million units) had never been so high, with General Motors by far the largest manufacturer, despite the failure of the Big Three to develop a 'world car.' Two years later a somewhat lower number of produced units represented 250 billion dollars. By the end of the 1980s the global car fleet amounted to 560 million units (nearly two-thirds of which were produced in the US, Germany, and Japan, and half of which were consumed there as well),[104] and another four decades later, when China had taken over the lead in world production, some estimates suggested that there were one billion motor vehicles plying the earth's roads. In terms of use, global travel volume (in passenger-kilometers) by the end of the last century had increased fivefold since 1960, the result of a tendency to 'hypermobility,' and increasing annual travel distance and population, the latter being the most influential factor in the long term.

But whereas local and transnational governments managed to curb local gaseous pollution, as we saw, CO_2 emissions continued to rise, with the world fleet of motor vehicles, calculated by the World Bank as 'producing' a quarter of a billion 'person-trips' per year, responsible for 12 percent (1988) of the greenhouse effect. The share of the US at that time in those 12 percent was still 38 percent, but the Global South's share was rapidly increasing. And although the problem of the car's role in the planet's climate change is one of the most urgent problems (see the next chapter), we should not forget that, from a mobility perspective, by far the most of the daily person-trips (more than ten billion) are made on foot (and half a billion on the bicycle).

In the previous chapters we have avoided giving an exact number (or share in the national populations) of the national middle classes (and especially the so-called 'new middle classes') that are supposed to be the principle buyers of these vehicles, whereas "the majority of women, children, the elderly and the disabled, as well as almost all poorer segments of the population in developing cities, are already practicing . . . sustainable forms of mobility," such as walking, cycling, and using public transport, formal and informal, the distinction between the two becoming increasingly blurred, the result being a product of "urban bricolage" and co-construction of rides.[105] Whether as the aggregate number of national estimates or as the 'global petty bourgeoisie' as defined by Giorgio Agamben, the estimates differ enormously; to give only two examples: Chinese estimates vary between 2 and 60 percent of the national population and the Indian middle classes were estimated to amount to between 25 and 100 million people. The uncertainty is further compounded by the phenomenon of a 'shrinking' or 'vanishing middle class' in the West as a result of neoliberal policies and the credit crisis of 2008 (most pronounced in the US and Sweden), on the one hand, and inflated prognostics by banks and other potential investors in the car fleet's growth in the Global South on the other.[106] Whereas the latter use a simple economic criterion (such as: those earning between \$2 and \$20 per day), scholars specializing in middle-class consumption, next to an economic measure, emphasize lifestyle criteria such as an urge to distinction, education for their children, engaging in leisure pursuits, and ownership of (first) a house and (then) a car. Hsin-Huang Michael Hsiao, working for the East Asia Middle Class Project of Taiwan's Academica Sinica, (re-)'discovered' the middle class in East Asia, and estimated it by the beginning of the 21st century as taking a share in the national population of between 30 and 60 percent. It is hard to see, however, how people earning between \$2 and, say, \$10 a day will ever be able to buy a car: we need much more detailed estimates to understand the current and future size of the car-buying middle class. If we take a more realistic income bracket of between $50 and $2,000 dollars per day, 1.2 billion earthlings can currently afford "luxury commodities like cars and travel."[107]

We also need more detailed studies about the intricacies of car-buying behavior in the Global South and East. The same applies to the transport and mobility policies of the local elites and their governments. One such vista can be gained from World Bank and IMF reports, accompanied by roadbuilders gathered in PIARC and the IRF, and the consultancy bureaus advising local elites about how to cope with the explosion of their domestic motorized mobilities. In the previous chapter we saw many examples of these local peculiarities, from the much higher share of transport in the household budget, spent on cars that often are many times more expensive than in the West, to the intensity of car use exposing drivers more than in the West to the dangers of road traffic. But for a global, *transnational* view we can look into two cases to get a first, better grip on this approach.

The first case is the clash between Western roadbuilding consultants and the realities in the Global South. In a dissertation from 1989, transport planner Harry Dimitriou interviewed sixty urban planners, many of whom conceded that they neglected the Southern dominance of non-motorized traffic, the preponderance of motorized two-wheelers, and public transport, while they never questioned the universality of their automotive orientation. "We over-built roads," one of them declared, and their proposals often contained solutions rejected in the consultants' own country.[108] *If* there exists a 'constant' in the enormous variety of local mobilities and its condition of 'automobile dependence' (as Australian critical sustainable transport professors Peter Newman and Jeff Kenworthy showed in 1989), it is *policy*.[109] This is an encouraging result, because it means that such dependencies can also be changed.

Another case is the road safety crisis, which was made global through the work of the WHO: in two reports (from 2004 and 2009) the organization redefined traffic safety in epidemiological terms, stressing the systemic violence of road traffic and pointing at alternatives, such as public transport. Since the start of automobilism, road traffic had demanded 60 million victims, quite an impressive figure if one compares this (if this usual reflex is permitted only once) with the 100 million war victims of the twentieth century. One of the researchers involved in WHO activities was professor Dinesh Mohan of Delhi's IIT university, who not only declared personal safety in traffic to be a 'human right,' but also showed that uncritically applying Western insights and safety measures did not work. For instance: enforced seatbelt use (one of the very effective measures in the West to bring down the number of fatalities among motorists) would in the heterogeneous traffic of the Indian capital lead to a decrease of only 1 percent. Gradually more local experts managed to contribute to the transnational debate (shown, for instance, by the publication of the *Journal of the Eastern Asia Society for Transportation Studies* from 2003, but also by the increasing readership from the Global South of the leading journal *Accident Analysis & Prevention* from 2004),[110] emphasizing the importance of the two-wheeler and informal transport.

Last, I cannot repeat often enough that, although the share of the world population involved in migration is now comparable to its share in the early twentieth century, it meanwhile dwindles compared to the half a billion annual tourists.[111] On a global scale carnivalesque sightseeing dominates, in a quantitative sense, the flows of migrants and refugees.

Notes

1. See for the following Gijs Mom, *Pacific Automobilism: Adventure, Status and the Carnival of Mobility, 1975–2015* (New York: Berghahn, 2022), passim.
2. Peter Freund and George Martin, "Driving South: The Globalization of Auto Consumption and Its Social Organization of Space," *Capitalism, Nature, Socialism* 11, no. 4 (December 2000): 51–71, here 55.
3. Edwin Black, *Internal Combustion: How Corporations and Governments Addicted the World to Oil and Derailed the Alternatives* (New York: St. Martin's Press, 2006), 200–1; Nicholas Dagen Bloom, *The Great American Transit Disaster: A Century of Austerity, Auto-Centric Planning, and White Flight* (Chicago: University of Chicago Press, 2023), 2–3; Katherine J. Goodwin, "Reconstructing Automobility: The Making and Breaking of Modern Transportation," *Global Environmental Politics* 10, no. 4 (November 2010): 60–78, here 67.
4. David Gartman, *Auto Opium: A Social History of American Automobile Design* (New York: Routledge, 1994), 156; David Gartman, "Three Ages of the Automobile: The Cultural Logics of the Car," *Theory, Culture & Society* 21, no. 4/5 (2004): 169–95, here 185; Shelley Nickles, "More is Better: Mass Consumption, Gender, and Class Identity in Postwar America," *American Quarterly* 54, no. 4 (December 2002): 581–622, here 583.
5. This paragraph is taken, with some adjustments, from Gijs Mom, *Globalizing Automobilism: Exuberance and the Emergence of Layered Mobility, 1900–1980* (New York: Berghahn, 2020), 37–38.
6. Immanuel Wallerstein, *World-Systems Analysis: An Introduction* (Durham, NC: Duke University Press, 2004), 77; Timothy Mitchell, *Carbon Democracy: Political Power in the Age of Oil* (London: Verso, 2013 [2011]), 193.
7. Steven Higashide, "Never Again Is Now: The Transportation Professions' Responsibility to Work Toward Justice," in Ryan Reft, Amanda K. Phillips de Lucas, and Rebecca C. Retzlaff (eds.), *Justice and the Interstates: The Racist Truth about Urban Highways* (Washington, DC: Island Press, 2023), 175–92, here 176.

8. A. Bruce-Briggs, *The War Against the Automobile* (New York: E.P. Dutton, 1977), quoted by Brian Ladd, *Autophobia: Love and Hate in the Automotive Age* (Chicago: University of Chicago Press, 2008), 146.
9. This section is based on the results of a book project funded by the Dutch Rijkswaterstaat and executed by transport engineer-turned-historian Ruud Filarski and this author. See Ruud Filarski and Gijs Mom, *Geschiedenis van de verkeersveiligheid in Nederland* [History of traffic safety in the Netherlands] (forthcoming 2025).
10. Robert D. Lifset, "A New Understanding of the American Energy Crisis of the 1970s," *Historical Social Research* 39, no. 4 (2014): 22–42, here 23 (italics in original).
11. Matthew J. Seelinger, "From the Jeep to the Humvee: U.S. Army Light Combat Vehicles, World War II to the Present," *On Point* 11, no. 3 (Winter 2005/2006): 8–13.
12. Christopher Neumaier, "Eco-Friendly versus Cancer Causing: Perceptions of Diesel Cars in West Germany and the United States 1970–1990," *Technology and Culture* 55, no. 2 (April 2014): 429–60. See for the following Gijs Mom, *The Electric Vehicle: Technology and Expectations in the Automobile Age* (Baltimore: Johns Hopkins University Press, 2004), 270–73.
13. Michel Callon, "The State and Technical Innovation: A Case Study of the Electrical Vehicle in France," *Research Policy* 9 (1980): 358–76.
14. Fernando J. Bosco, "Actor-Network Theory, Networks, and Relational Approaches in Human Geography," in Stuart Aitken and Gill Valentine (eds.), *Approaches to Human Geography* (London: Sage, 2006), 136–46, here 136.
15. See for instance Tim Ingold, "When ANT Meets SPIDER: Social Theory for Arthropods," in C. Knappett and L. Malafouris (eds.), *Material Agency: Towards a Non-Anthropocentric Approach* (New York: Springer, 2008), 209–15, who ends a dialogue between an 'ant' and a 'spider' by having the latter (representing a form of skilled agency) exclaim: "I cannot, for the most part, understand a word of what you say" (ibid., 215); "eliding": John Law, "After ANT: Complexity, Naming and Topology," in John Law and John Hassard (eds.), *Actor Network Theory and After* (Oxford/Malden, MA: Blackwell/*Sociological Review*, 1999), 1–14, here 1.
16. Matthias Wieser, *Das Netzwerk von Bruno Latour: Die Akteur-Netzwerk-Theorie zwischen Science & Technology Studies and poststrukturalistischer Soziologie* (Bielefeld: transcript, 2012), 81.
17. Michael Shnayerson, *The Car That Could: The Inside Story of GM's Revolutionary Electric Vehicle* (New York: Random House, 1996), 274n18.
18. David A. Kirsch, *The Electric Vehicle and the Burden of History* (New Brunswick, NJ: Rutgers University Press, 2000).
19. Luis F. Alvarez León and Yuko Aoyama, "Industry Emergence and Market Capture: The Rise of Autonomous Vehicles," *Technological Forecasting & Social Change* 180 (2022): 121661, 3.
20. Greg Perkins and Johann Peter Murmann, "What Does the Success of Tesla Mean for the Future Dynamics in the Global Automobile Sector?" *Management and Organization Review* 14, no. 3 (September 2018): 471–80; market share: Matteo Vivi and Anne-Mette Hermans, "'Zero Emission, Zero Compromises': An Intersectional, Qualitative Exploration of Masculinities in Tesla's Consumer Stories," *Men and Masculinities* 25, no. 4 (2022): 622–44, here 623.
21. See for instance Frank W. Geels, Geoff Dudley, and René Kemp, "Findings, Conclusions and Assessments of Sustainability Transitions in Automobility," in Frank W. Geels, René Kemp, Geoff Dudley, and Glenn Lyons (eds.), *Automobility in*

Transition? A Socio-Technical Analysis of Sustainable Transport (New York: Routledge, 2012), 335–73, here 343; Prius statistics and fleets: Renato J. Orsato, Marc Dijk, René Kemp, and Masaru Yarime, "The Electrification of Automobility: The Bumpy Ride of Electric Vehicles toward Regime Transition," in ibid., 205–28, here 213, 217; fifteen million: Bryan Appleyard, *The Car: The Rise and Fall of the Machine That Made the Modern World* (London: Weidenfeld & Nicolson, 2022), 255.

22. Lance Noel, Benjamin K. Sovacool, Johannes Kester, and Gerardo Zarazua de Rubens, "Conspicuous Diffusion: Theorizing How Status Drives Innovation in Electric Mobility," *Environmental Innovation and Social Transitions* 31 (2019): 154–69, here 163–64; 2012 and 2018: Tomas Moe Skjølsvold and Marianne Ryghaug, "Temporal Echoes and Cross-Geography Policy Effects: Multiple Levels of Transition Governance and the Electric Vehicle Breakthrough," *Environmental Innovation and Social Transitions* 35 (2020): 232–40, here 234.
23. Benjamin Sovacool, Wokje Abrahamse, Long Zhang, and Jingzheng Ren, "Pleasure or Profit? Surveying the Purchasing Intentions of Potential Electric Vehicle Adopters in China," *Transportation Research Part A* 124 (2019): 69–81, here 70 (Figure 1), 77; according to Pablo Ampuero-Ruiz (University of Amsterdam) local emission-free use does not play a role in purchasing motives of Chinese EV buyers ("How the Adoption of Electromobility is Redefining the Car in China," oral presentation at the conference Cultures of Automobility, University of Konstanz, 10–12 October 2024).
24. The following paragraphs are taken, with some adjustments, from Mom, *Pacific Automobilism*, 5–6; Jed Esty and Colleen Lye, "Peripheral Realisms Now," *Modern Language Quarterly* 73, no. 3 (September 2012): 269–88.
25. Lukas Hoffmann, *Postirony: The Nonfiction Literature of David Foster Wallace and Dave Eggers* (Bielefeld: transcript, 2016), 370.
26. Tim Dant and Peter J. Martin, "By Car: Carrying Modern Society," in Jukka Gronow and Alan Warde (eds.), *Ordinary Consumption* (London: Routledge, 2001), 143–57, here 143; Sven Kesselring and Gerlinde Vogl, "The New Mobilities Regime," in Susanne Witzgall, Gerlinde Vogl, and Sven Kesselring (eds.), *New Mobilities Regimes in Art and Social Sciences* (Farnham: Ashgate, 2013), 17–36, here 28.
27. Leo Marx, *The Pilot and the Passenger: Essays on Literature, Technology, and the Culture in the United States* (New York: Oxford University Press, 1988).
28. Eric Laurier et al., "Driving and 'Passengering': Notes on the Ordinary Organization of Car Travel," *Mobilities* 3, no. 1 (March 2008): 1–23. The interest in social science circles can be fathomed from the extraordinarily large number of coauthors (fourteen) of this article.
29. Dana A. Heller, *The Feminization of Quest-Romance: Radical Departures* (Austin: University of Texas Press, 1990), 10, 12.
30. Ginette Verstraete, *Tracking Europe: Mobility, Diaspora, and the Politics of Location* (Durham, NC: Duke University Press, 2010), 87.
31. Mariusz Jastrząb, "Cars as Favors in People's Poland," in Lewis H. Siegelbaum (ed.), *The Socialist Car: Automobility in the Eastern Bloc* (Ithaca, NY: Cornell University Press, 2011), 30–46, here 31 (mass motorization).
32. Lewis H. Siegelbaum, "The Impact of Motorization on Soviet Society after 1945," in Manfred Krieger, Ulrike Gutzmann, and Dirk Schlinkert (eds.), *Towards Mobility: Varieties of Automobilism in East and West* (Wolfsburg: Volkswagen AG, 2009), 21–29, here 22.

33. Jukka Gronow and Sergei Zhuravlev, "Soviet Luxuries from Champagne to Private Cars," in David Crowley and Susan E. Reid (eds.), *Pleasures in Socialism: Leisure and Luxury in the Eastern Bloc* (Evanston, IL: Northwestern University Press, 2010), 121–46, here 135, 140; one-third: Lewis H. Siegelbaum, "On the Side: Car Culture in the USSR, 1960s–1980s," *Technology and Culture* 50, no. 1 (January 2009): 1–23, here 20.
34. Lela Rekhviashvili and Wladimir Sgibnev, "Theorising Informality and Social Embeddedness for the Study of Informal Transport: Lessons from the *Marshrutka* Mobility Phenomenon," *Journal of Transport Geography* 88 (2020): 102386, 2; perestroika and immigrants: Anna Sanina, "The *Marshrutka* as a Socio-Cultural Phenomenon of a Russian Megacity," *City, Culture and Society* 2 (2011): 211–18, here 211.
35. Mark Keck-Szajbel, "Hitchikers' Paradise: The Intersection of Mass Mobility, Consumer Demand, and Ideology in the People's Republic of Poland," in Cathleen M. Giustino, Catherine J. Plum, and Alexander Vari (eds.), *Socialist Escapes: Breaking Away from Ideology and Everyday Routine in Eastern Europe, 1945–1989* (New York: Berghahn, 2013), 167–86, here 168, 178.
36. Harley Balzer, "Russia's Middle Classes," *Post-Soviet Affairs* 14, no. 2 (1998): 165–86, here 166 (engineers), 172 (15 percent), 178 (car ownership); Anna Kruglova, "Driving in Terrain: Automobility, Modernity, and the Politics of Statelessness in Russia," *American Ethnologist* 46, no. 4 (2019): 457–69, here 459, 461 (italics in original).
37. Jennifer Patico, "To Be Happy in a Mercedes: Tropes of Value and Ambivalent Visions of Marketization," *American Ethnologist* 32, no. 3 (2005): 479–96, here 490.
38. Martin Müller, "In Search of the Global East: Thinking between North and South," *Geopolitics* 25, no. 3 (2020): 734–55, here 734, 736.
39. Egor Muleev, "*Skotovoz, pripiski* and the 'Law of Labour Settlement': Transportation Planning and Management in the USSR," *Journal of Transport History* 45, no. 1 (2024): 62–82, here 73 ("gap"); Egor Muleev, "Why do Marshrutkas Exist in One City and Not in Others? Toward a Political Economy of Routes in Russian Urban Public Transportation," *Journal of Economic Sociology* 21, no. 2 (March 2020): 99–114, here 100 (inconsistent), 101 ("production"); Lela Rekhviashvili, "We Need to Talk about Political Society: Subaltern Resistances beyond Civil Society in Eastern Europe and Eurasia," *Central Asian Survey* 42, no. 2 (2023): 219–37, here 225.
40. Giovanni Arrighi, *The Long Twentieth Century: Money, Power, and the Origins of Our Times* (London: Verso, 1994), 350.
41. Quoted in Jason Henderson, "The Politics of Mobility: De-Essentializing Automobility and Contesting Urban Space," in Jim Conley and Arlene Tigar McLaren (eds.), *Car Troubles: Critical Studies of Automobility and Auto-Mobility* (Farnham: Ashgate, 2009), 147–64, here 148. Iraq: see "Blackwater in Iraq" by Harper's Magazine—YouTube (last consulted on 3 November 2024).
42. George Baca, "Neoliberal Narratives of Crisis: The Feeble Cries of a Vanishing 'Class,'" *Dialectical Anthropology* 41 (2017): 377–85, here 378.
43. Mimi Sheller and John Urry, "The New Mobilities Paradigm," *Environment and Planning A* 38 (2006): 207–26.
44. Zygmunt Bauman, *Liquid Life* (Cambridge: Polity Press, 2006 [2005]); outside: Pauline Wolff, "From the Urban Planning Discourse to a Circulation *Dispositif*: An Epistemological Approach to the Mobility Turn," in Marcel Endres, Katha-

rina Manderscheid, and Christophe Mincke (eds.), *The Mobilities Paradigm: Discourses and Ideologies* (London: Routledge, 2016), 202–23, here 205.

45. Rae Francis C. Quilantang and Kimberly Nicole Panis Quilantang, "'Para po! (Stop, Please)': The Cultural Significance of Jeepney Decorations in the Philippines in the Context of Modernisation," *Mobility Humanities* 3, no. 1 (January 2024): 133–48, here 136.
46. Mimi Sheller, "Pluralizing Mobilities Theory for Post-Carbon Futures and Social Justice," *Transfers* 13, no. 1/2 (Spring/Summer 2023): 162–77; last quote: Janet Elspeth McAllister, "Stuck Fast: A Critical Analysis of the 'New Mobility Paradigm,'" master's thesis (Auckland: University of Auckland, 2011), 9.
47. Slavoj Žižek, *Violence: Six Sideways Reflections* (New York: Picador, 2008), 29.
48. Dag Balkmar, "On Men and Cars: An Ethnographic Study of Gendered, Risky and Dangerous Relations," unpublished doctoral dissertation (Linköping: Linköping University, 2012).
49. Christine L. Lezotte, "Have You Heard the One about the Woman Driver? Chicks, Muscle, Pickups, and the Reimagining of the Woman behind the Wheel," unpublished doctoral dissertation (Bowling Green, OH: Bowling Green State University, May 2015).
50. Quoted in Katie Mills, *The Road Story and the Rebel: Moving through Film, Fiction, and Television* (Carbondale: Southern Illinois University Press, 2006), 146.
51. Pascal Menoret, *Joyriding in Riyadh: Oil, Urbanism, and Road Revolt* (New York: Cambridge University Press, 2014).
52. Jenny Sundén, "A Sense of Play: Affect, Emotion and Embodiment in *World of Warcraft*," in Marianne Liljeström and Susanna Paasonen (eds.), *Working with Affect in Feminist Readings: Disturbing Differences* (London: Routledge, 2010), 45–57.
53. Joseph S. Nye Jr., *Soft Power: The Means to Success in World Politics* (New York: Public Affairs, 2004).
54. Chris Bramall, *Chinese Economic Development* (London: Routledge, 2009); Leslie T. Chang, *Factory Girls: From Village to City in a Changing China* (New York: Spiegel & Grau, 2008).
55. Stevan Harrell, *An Ecological History of Modern China* (Seattle: University of Washington Press, 2023), 167, 182, 411–21.
56. Han Sang Kim, "'My Car' Modernity: What the U.S. Army Brought to South Korean Cinematic Imagination about Modern Mobility," *Journal of Asian Studies* 75, no. 1 (February 2016): 63–85, here 63.
57. Vladimir Popov, "The Culture of New Mobility in Russia: Networks and Flows Formation," *Mobilities* 7, no. 1 (February 2012): 151–69, here 157.
58. Ravi Arvind Palat, "Miracles of the Day Before? The Great Asian Meltdown and the Changing World-Economy," *Development and Society* 28, no. 1 (July 1999): 1–47, here 28.
59. See, also for the following, Mom, *Pacific Automobilism*, 712–17.
60. Thomas J. Campanella, *The Concrete Dragon: China's Urban Revolution and What It Means for the World* (New York: Princeton Architectural Press, 2008), 151.
61. Yajun Mo, *Touring China: A History of Travel Culture, 1912–1949* (Ithaca, NY: Cornell University Press, 2021), ix.
62. Wendy Dobson, *Living with China: A Middle Power Finds Its Way* (Toronto: University of Toronto Press, 2019), 93 (trillion; the quote stems from the World Bank); Srinivasa Madhur, Ganeshan Wignaraha, and Peter Darjes, *Roads for Asian Integration: Measuring ADB's Contribution to the Asian Highway Network*

(n.p.: Asian Development Bank, November 2009), 17 (Table 5). See 16 for a map of the network.

63. Lela Rekhviashvili and Thilo Lang, "Chinese Investments as Part of Infrastructure-Led Development: Multi-Scalar Contestations around Georgia's Flagship Infrastructure Projects," *Eurasian Geography and Economics* (7 February 2024): 5, https://doi.org/10.1080/15387216.2024.2311712.
64. Seth Schindler, Jessica DiCarlo, and Dinesh Paudel, "The New Cold War and the Rise of the 21st-Century Infrastructure State," *Transactions of the Institute of British Geographers* 47 (2022): 331–46, here 221, 335, 338, 340.
65. Luigi Tomba, *The Government Next Door: Neighborhood Politics in Urban China* (Ithaca, NY: Cornell University Press, 2014), 152.
66. Li Chunling, "Profile of China's Middle Class," in Li Chunling (ed.), *The Rising Middle Classes in China* (n.p.: Paths International/Social Sciences Academic Press [China], 2012), 87–125, here 124: Li estimated the middle class's size at 13 percent of the working population.
67. Eva Chen, "Fashioning the Cosmopolitan Girl: Sartorial Display and Technologies of Femininity in the Chinese Bestseller *Du Lala* (2007–11)," *Fashion Theory* 17, no. 5 (2013): 553–77, here 558, 565.
68. Lingjing Zhan and Yanqun He, "Understanding Luxury Consumption in China: Consumer Perceptions of Best-Known Brands," *Journal of Business Research* 65 (2012): 1452–60, here 1458.
69. Wei Hui [Zhou Weihui], *Shanghai Baby* (London: Constable, 2002), 3, 46.
70. Eva Chen, "Shanghai(ed) Babies: Geopolitics, Biopolitics and the Global Chick Lit," *Feminist Media Studies* 12, no. 2 (2012): 214–28, here 217; heterosexual, quasi-feminist: Eva Yin-I Chen, "*Shanghai Baby* as a Chinese Chick-Lit: Female Empowerment and Neoliberal Consumerist Agency," *Asian Journal of Women's Studies* 15, no. 1 (2009): 54–93, here 62, 74.
71. For a rare ethnographic investigation of car purchase and use see Jun Zhang, *Driving toward Modernity: Cars and the Lives of the Middle Class in Contemporary China* (Ithaca, NY: Cornell University Press, 2019).
72. Arve Hansen, "*Doi moi* on Two and Four Wheels: Capitalist Development and Motorised Mobility in Vietnam," in Arve Hansen and Kenneth Bo Nielsen (eds.), *Cars, Automobility and Development in Asia: Wheels of Change* (London: Routledge, 2017), 103–19, here 114.
73. Denise Potrzeba Lett, *In Pursuit of Status: The Making of South Korea's "New" Urban Middle Class* (Cambridge, MA: Harvard University Press, 1998); Seung-Kuk Kim, "Changing Lifestyles and Consumption Patterns of the South Korean Middle Class and New Generations," in Chua Beng-Huat (ed.), *Consumption in Asia: Lifestyles and Identities* (London: Routledge, 2000), 61–81.
74. M. William Steele, "The Making of a Bicycle Nation: Japan," *Transfers* 2, no. 2 (Summer 2012): 70–94, here 80–84 (quote on 83); third: M. William Steele, "The Speedy Feet of the Nation: Bicycles and Everyday Mobility in Modern Japan," *Journal of Transport History* 31, no. 2 (December 2010): 182–209, here 203.
75. Xiao Lin, Peter Wells and Benjamin Sovacool, "The Death of a Transport Regime? the Future of Electric Bicycles and Transportation Pathways for Sustainable Mobility in China," *Technology Forecasting & Social Change* 132 (2018) 255–267. The terms "electric bicycles" (in the title) and "e-bike" (in the text) are somewhat misleading, as the statistics given also include "e-scooters" (ibid., 255), and probably also electric motorcycles. 80 percent of car drivers possess such an electric two-wheeler as second vehicle (ibid., 263).

76. Pavan K. Varma, *The Great Indian Middle Class* (New Delhi: Penguin, 2007), 174, 184 ("opulence").
77. Rameshwar Dayal Sharma, Sandeep Jain, and Kewal Singh, "Growth Rate of Motor Vehicles in India—Impact of Demographic and Economic Development," *Journal of Economic and Social Studies* 1, no. 2 (July 2011): 137–53, here 138 and 147 (Table 2).
78. Geetam Tiwari, "Low Cost Means of Transport in Cities: The Critical Element in City Transport System in Low Income Countries," in X. Godard and I. Fatonzoun (eds.), *Urban Mobility for All/La Mobilité Urbaine pour Tous: Proceedings of the Tenth International CODATU Conference, Lomé, Togo, 12–15 November 2002* (Lisse: A. A. Balkema, 2002), 215–20, here 215 (5–20 percent); Rajat Bose and P. K. Sarkar, "The Role of Cycle Rickshaws—Low Cost Means of Transport in Old Part of a City," in ibid., 199–203, here 201; Lake Sagaris and Anvita Arora, "Cycling for Social Justice in Democratizing Contexts: Rethinking 'Sustainable' Mobilities," in Tanu Priya Uteng and Karen Lucas (eds.), *Urban Mobilities in the Global South* (London: Routledge, 2017), 19–40, here 30 (600,000).
79. Gaurav Mittal, "The State and the Production of Informalities in Urban Transport: *Vikrams* in Dehradun, India," *Geoforum* 136 (2022): 273–82, here 278.
80. Annemiek Prins, "Unfixing the City: Rickshaw Mobilities, Modernities and Urban Change in Dhaka," unpublished dissertation (Aberdeen: University of Aberdeen, 2020), 85, 103n52 (Knowles quoted on 75).
81. Jonas van der Straeten, "Sustainability's 'Other': Coming to Terms with the Electric Rickshaw in Bangladesh," *Historical Social Research* 47, no. 4 (2022): 139–67, here 147, 149, 151, 153.
82. Dennis Zuev, David Tyfield, and John Urry, "Where is the Politics? E-Bike Mobility in Urban China and Civilizational Government," *Environmental Innovation and Societal Transitions* 30 (2019): 19–32.
83. Govind Gopakumar, *Installing Automobility: Emerging Politics of Mobility and Streets in Indian Cities* (Cambridge, MA: MIT Press, 2020); Ritajyoti Bandyopadhyay, *Streets in Motion: The Making of Infrastructure, Property, and Political Culture in Twentieth-Century Calcutta* (Cambridge: Cambridge University Press, 2022); Tarini Bedi, *Mumbai Taximen: Autobiographies and Automobilities in India* (Seattle: University of Washington Press, 2022).
84. Prema-Chandra Athukorala and C. Veeramani, "From Import Substitution to Integration into Global Production Networks: The Case of the Indian Automobile Industry," *Asian Development Review* 36, no. 2 (2019): 72–99, here 74.
85. Yasmin Hussain, *Writing Diaspora: South Asian Women, Culture and Ethnicity* (Aldershot: Ashgate, 2005), 97.
86. John Howe, "Sustaining Africa's Rural Networks: The Asset Management Approach," *World Transport Policy & Practice* 5 (1999): 11–16, here 11.
87. Kenda Mutongi, *Matatu: A History of Popular Transportation in Nairobi* (Chicago: University of Chicago Press, 2017), 95, 34.
88. Robert Heinze, "Fighting over Urban Space: Matatu Infrastructure and Bus Stations in Nairobi, 1960–2000," *Africa Today* 65, no. 2 (Winter 2018): 2–21, here 14–16.
89. Martin Oteng-Ababio and Ernest Agyemang, "Virtue out of Necessity? Urbanisation, Urban Growth and Okada Services in Accra, Ghana," *Journal of Geography and Geology* 4, no. 1 (March 2012): 148–62, here 149; Martin Oteng-Ababio and Ernest Agyemang, "The Okada War in Urban Ghana: A Polemic Issue or Policy Mismatch?" *African Studies Quarterly* 15, no. 4 (September 2015): 25–44, here

28, 31; deregulation: Davison Muchadenyika, "Informal Transport, Politics and Power in Harare," in Daniel E. Agbiboa (ed.), *Transport, Transgression and Politics in African Cities: The Rhythm of Chaos* (Abingdon: Routledge, 2019), 46–77, here 65; mafia: Daniel E. Agbiboa, "Stomach Infrastructure: Informal Transport, Electoral Politics and the Precariousness of Patronage in Lagos," in ibid., 43–59, here 65; labor: Robert Heinze, "'Taxi Pirates': A Comparative History of Informal Transport in Nairobi and Kinshasa, 1960s–2000s," in ibid., 19–39, here 19; attack and commuting: Daniel E. Agbiboa, "State Law as a Means of Resistance: Okada Riders versus the Lagos State Government," in ibid., 147–71, here 157, 161.

90. Matteo Rizzo, "Being Taken for a Ride: Privatisation of the Dar Es Salaam Transport System 1983–1998," *Journal of Modern African Studies* 40, no. 1 (2002): 133–57, here 144.
91. Camilla Ida Ravnbøl, Trine Mygind Korsby, and Anja Simonsen, "Introduction: Transnational Street Business; Migrants in the Informal Economy," *Migration and Society* 6 (2023): 1–15, here 2.
92. Marie Richard Zouhoula Bi, "The Woro-Woro *Gares Routières* of Abidjan: Artisanal Transport and Local Governance in Côte d'Ivoire's Largest City," *Africa Today* 65, no. 2 (Winter 2018): 22–33, here 32.
93. Robert Cervero, *Informal Transport in the Developing World* (Nairobi: United Nations Centre for Human Settlements [Habitat], 2000), 30, 68.
94. Mia Bay, *Traveling Black: A Story of Race and Resistance* (Cambridge, MA: Belknap Press of Harvard University Press, 2021), 161.
95. Peter Chilson, *Riding the Demon: On the Road in West Africa* (Athens: University of Georgia Press, 1999), 15; Neil Carrier, "'*Miraa* is Cool': The Cultural Importance of *Miraa* (Khat) for Tigania and Igembe Youth in Kenya," *Journal of African Cultural Studies* 17, no. 2 (December 2015): 201–18, here 202.
96. Rizzo, "Being Taken for a Ride," passim.
97. Joshua Grace, *African Motors: Technology, Gender, and the History of Development* (Durham, NC: Duke University Press, 2021), 146.
98. Emily Callaci, "Street Textuality: Socialism, Masculinity, and Urban Belonging in Tanzania's Pulp Fiction Publishing Industry, 1975–1985," *Comparative Studies in Society and History* 59, no. 1 (2017): 183–210, here 201.
99. Anna-Leena Toivanen, *Mobilities and Cosmopolitanisms in African and Afrodiasporic Literatures* (Leiden/Boston: Brill/Rodopi, 2021), 128n6 ('luck'), 129 (nightclub), 129n7 (prostitute), 132 (tourists), 133 (flashbacks), 140 (Mbembe; the quote is Toivanen's paraphrase).
100. Catherine Mazauric, *Mobilités d'Afrique en Europe: Récits et figures de l'aventure* (Paris: Karthala, 2012), 50.
101. Eduardo A. Vasconcellos, "Reassessing Traffic Accidents in Developing Countries," *Transport Policy* 2, no. 4 (1996): 263–69, here 267.
102. Roberto Bolaño, *The Savage Detectives* (London: Picador, 2009 [1998]), 533.
103. Robert Braun and Richard Randell, *Post-Automobility Futures: Technology, Power, and Imaginaries* (Lanham, MD: Rowman and Littlefield, 2022), 100 (CIA); *Infrastructure at the Crossroads: Lessons from 20 Years of World Bank Experience* (Washington, DC: World Bank, 2006), xiii.
104. George Martin, "The Global Intensification of Motorization and Its Impacts on Urban Social Ecologies," in Conley and McLaren (eds.), *Car Troubles*, 219–33, here 220.

105. Tanu Priya Uteng and Karen Lucas, "The Trajectories of Urban Mobilities in the Global South: An Introduction," in Tanu Priya Uteng and Karen Lucas (eds.), *Urban Mobilities in the Global South*, 1–18, here 14; blurred and "bricolage": Dirk Heinrichs, Daniel Ehebrecht, and Barbara Lenz, "Moving beyond Informality? Theory and Reality of Public Transport in Urban Africa," in ibid., 134–54, here 134.
106. The following is primarily based on Göran Therborn, "Dreams and Nightmares of the World's Middle Classes," *New Left Review* 124 (July–August 2020): 63–87.
107. Oliver Schöller-Schwedes and Stephan Rammler, *Mobile Cities: Dynamiken weltweiter Stadt- und Verkehrsentwicklung* (Berlin: LIT, 2008), 91.
108. Harry T. Dimitriou, "A Developmental Approach to Urban Transport Planning: The Urban Transport Planning Process and Its Derivatives; An Assessment of their Contribution to Transport Planning for Third World Cities; Volume I," doctoral dissertation (Cardiff: University of Wales, October 1989), 80, 166.
109. Peter W. G. Newman and Jeffrey R. Kenworthy, *Cities and Automobile Dependence: A Sourcebook* (Aldershot: Avebury Technical, 1992 [1989]).
110. Xin Zou, Hai L. Vu, and Helai Huang, "Fifty Years of Accident Analysis & Prevention: A Bibliometric and Scientometric Overview," *Accident Analysis & Prevention* 144 (2020): 1–26, here 5 (Figure 7).
111. Adam McKeown, "Global Migration, 1846–1940," *Journal of World History* 15, no. 2 (2004): 155–89, here 185 (the statistics are from 1990).

CHAPTER 5

The End of the Car and Its Alternatives

> How . . . to relate to the future without capturing it and neutralizing it before it happens?
>
> —Ben Anderson and Paul Harrison, "The Promise of Non-Representational Theories"

Now that we have completed our historical narrative it is time to look around (instead of back) and formulate where we are and try to 'glance' in a historically informed way into possible futures of automobilism (and beyond). Looking around in 2010, Katherine J. Goodwin observed: "If taken seriously, tackling climate change will require most of us in the industrialized world to eat differently, move differently, and think about humanity's place in the world differently."[1] Scholars, politicians, and activists from many different backgrounds have meanwhile formulated strategies to bring such change about, and not only in the West.

Changing the current automotive system presupposes a thorough knowledge of its history: as we have seen in the previous chapters, on a planetary scale, we observe two distinctive, but entangled, global subcultures of automobilism, one Western, adventurous, either tamed or over-the-top, absent-present or carnivalesque, the other beyond the West where the car (as a vehicle of status rather than as an 'adventure machine') is overwhelmed by other motorized and non-motorized mobilities, most particularly two-wheelers (including rickshaws) and so-called jitneys. This is not so much a distinction of vehicles; rather, the vehicles function as a *pars pro toto* for cultures, suggesting a difference between an individualized and a collectivist culture (as can be taken from belletristic and related utterances), the border between these cultures slowly dissolving, as if 'world road motorization' is converging into a planetary culture.

Investigating the current alternatives available, or promised to be ready in the (near) future, we may begin to ask, which alternatives were and are offered by the *industry*? In Chapters 1 and 4 we have followed the first three generations and the 1970s generation of the *electric pas-*

senger car, an alternative that for most early engineers and other specialists of taxi and rental fleets and some car manufacturers formed a better (more reliable, quieter, accessible to women) urban alternative to the combustion engine car, but both the industry and the majority of motorists (most of them men) were much more reluctant to consider an option that threatened to jeopardize an open (and later more covert) adventurous culture of violence and aggression.

This is also a personal story: the last chapters in my dissertation (1997) on the history of the electric vehicle (EV) were inspired, when I was a young automotive engineer emancipating into a historian of technology,[2] by the thrilling variety of proposed solutions by small start-ups and individual researchers in 1990s California and the simultaneous realization that the car industry was not really interested. Thrilling too in a corporeal sense: during a test drive in Stanford with the General Motors Impact (the proto-version of the notorious EV1, see Chapter 4) I left all combustion cars far behind when accelerating from a traffic light. In California, a vanguard in applying the national Clean Air Act Amendments of 1990, I visited electrical and mechanical engineers made redundant in the local defense industry, who developed their hybrids in garages in Stanford, combining mini-gas turbines with novel battery technologies and other 'adventurous' combinations.

In La Rochelle, I witnessed the start of a fleet test of EVs made available by French car manufacturers and the national electric utility (who sought outlets for their impending overcapacity of nuclear energy), and in Germany and the Netherlands I visited teams of engineers in research institutes and the electric utility industry busy developing 'high-energy batteries' (nickel-metal hydride, sodium-sulfur combinations) and the 'fuel cell' (which works on hydrogen and promises to be virtually emission-free). New, electronically controlled electric motors were developed, some of them mounted in the wheels; in general, a vibrant atmosphere of innovation was in the air, inspiring engineers and social scientists alike.

At the same time I saw Transition Studies emerge, not much hindered by knowledge of automotive technology and its history, following the whims of the industry, now a hybrid hype, then a full battery-electric vehicle hype.[3]

[[Transition Studies started in the Netherlands as a so-called 'multilevel perspective' within Sustainable Transitions, based on the idea that radical (socio)technical innovations emerge in 'niches' that have to be protected against counterforces from higher levels.[4] Transition Studies thus is an inherently performative field (initially also called 'Strategic Niche Management,' SNM) and appeared very attractive to policymakers. From the start in the 1980s transition students often used transport historians' products to develop strongly patterned, rather "poor" case studies, but when it became very successful internationally other topics and more sophisticated historical case studies necessitated a constant adjustment ("refinement") to its theory.[5] When entering into transport studies as an approach "beyond technology fix or behaviour change,"[6] however, it encountered an already established New Mobilities paradigm as a rival approach, criticizing its technical bias, its errors as nonspecialists, and suggesting that nontechnical niches also exist, as not all change emanates from the bottom up.[7] Ironically, the recent breakthrough of EVs is precisely such a phenomenon, pushed through by national and transnational governments under pressure from a security crisis and from municipal governments trying to abate the urban crisis.]]

Against the constantly repeated cycle of hope and disappointment, I assumed, somewhat provocatively, that perhaps the EV would forever remain a vehicle of tomorrow, unless local authorities, using a big crisis, broke the spell of the gas car by simply forcing manufacturers to bring 'zero emission vehicles' (ZEVs) onto the market, as for a brief period seemed to happen with the California Air Resources Board's (CARB) 'mandate' of the late 1990s. Before I entered into the field of academic history I cowrote a small book on my experiences and insights, calling the EV a Trojan horse, which, based on my evolutionary argumentation (see Chapter 1), I saw sneak into car technology as an added electric motor in a hybrid, slowly expanding and overtaking the entire propulsion system, as an organ growing wild.[8]

And then, a real crisis (and a widespread fear of more crises) occurred, triggered by Russia's invasion of Ukraine, an energy supply crisis felt by entire national populations, forcing their governments to intervene, further incited by the perspective of a 'leapfrogging' China, and supported by an ever better informed and politically more mature anti-climate change activism. It brings back to mind the energy crises of the 1970s, which "enabled Washington and the oil companies to move to a system of higher energy prices."[9] The climate conference in Dubai in 2023 resulted in a "UEA consensus" among 197 countries to phase out the fossil fuel car. Equally nonbinding is a resolution prepared by US Congress members to start out-phasing combustion cars from 2030 and have a fully fossil-free transport system by 2050.[10] As we saw in the previous chapter, Norway was the first in 2017 to announce a phase-out of internal combustion cars, demanding the exclusive sale of 'zero emission' cars in 2025, soon followed by Denmark, Iceland, Ireland, the Netherlands, Slovenia, and Sweden (all aiming at 2030), in 2020 followed by the UK (aiming at 2035; Scotland: 2032), and France and Spain (aiming at 2040).[11]

The future of the car seemed more and more irreversibly aimed at an electric future when the United Nations Climate Change Conference of 2021 in Glasgow ushered in a (nonbinding) call to sell zero emission vehicles worldwide by 2040 in an effort to keep the world's temperature rise below 1.5 degrees.[12] When publicist Bryan Appleyard, author of *Aliens: Why They Are Here*, was interviewed by American news website for 'explanatory journalism' *Vox* about his recent book on "the rise and fall" of "the car as we know it," he warned against Silicon Valley "tak[ing] over now . . . They're doing this to grab another source of information" with the result that "the pleasure of the car itself and the internal combustion engine are going to be left behind." In his book, written "to celebrate the drama and beauty of what is soon to be lost," he connected "the last days of the car" explicitly with America's "decline."[13] Popular car literature (mobility populism) is replete with such paranoid comments.[14]

This time, the car industry was remarkably silent about the 'U-turn,' although some pleas for more 'objectivity' could be heard from related research institutions.[15] The industry produced hardly any controversial LCAs (life-cycle assessments) and seemed to agree, facing the Chinese threat of cheap EVs, with the conversion, counting on governmental support during the operation. Apart from a rare assessment of diesel cars being cleaner than EVs and the production of EVs being less sustainable than that of the combustion version, most experts now believe that even with the coal-dominated energy mix of China, electric propulsion is still more beneficial to the climate than fossil fuels, although criticism of the unreliability of LCAs in general remain.[16]

Yet geographer Jason Henderson convincingly argued, based on a review of scholarly journals from 2018 and 2019, that "EV production and consumption might escalate rather than reduce global resource and energy demand," pointing at the same time at the "carbon gentrification" in the West driven by the assumption that the break-even point of EVs in LCA comparisons is only so low because EVs are considered to have to cover a larger distance than their fossil rivals, a fact that partially explains the previously mentioned silence and cooperative attitude of the industry.[17] Preferring what he calls "green mobilities such as cycle lanes, bus lanes, and pedestrianization of cities," he warned that EVs will "compete [with] or hinder" such developments, also literally, as competitors for scarce urban space, especially in the aimed-for compact, post-suburban cities.[18] The idea of "more cars and more driving as the fundamental assumption in EV narratives" combined with the insight that "conventional gasoline cars remain for many decades before a full transition to EVs," while the preference for electrifying SUVs increases the "new car CO_2" and "kinetic underclasses" will be excluded from both car types, leads to the conclusion that this solution will arrive (if at all) much too late for the crucial contribution from transportation in the implementation of the climate agreements of Paris (2015) and Glasgow. In those cities, meetings of tens of thousands of participants took place including young and very young activists who support the voicing of the interests of the Global South, despite their conclusion that the process is "captured by corporate forces."[19]

The compliant attitude of the industry may be related to the fact that the EV has been coupled by many observers with the perspective of a fully automated, 'driverless' function, a perspective that opens a full-blown 'automotive arms race' with China. Although part of the 'fossil fuel paradigm' (just like the EV, as a vehicle of tomorrow for a century), so-called 'autonomous vehicles' (AVs) have been linked to electric propulsion by prominent EV proponents.[20] Amid a plethora of overly optimistic assessments of AVs' future, by far the best *historical* approach is given in Fabian Kröger's analysis, which places the origins of this propulsion-cum-management system squarely in the 'fossil' 1950s (and it probably originated in airplanes), although as a utopian as well as dystopian fantasy it is as old as the car, especially after wireless mobile radio was introduced as part of automotive technology. By the end of the 1980s, when the development of in-car electronics experienced an acceleration, fourteen European car manufacturers and eleven European countries started a seven-year research program of nearly half a billion dollars called Prometheus aimed at developing an 'automated car' guided by cables in the road or by radio beacons, in order to catch up with the US and Japan, where similar projects were initiated. The project failed because of liability concerns by the industry. In France, the Praxitèle project during the 1990s added a third innovation (next to electric driving and autonomy): the integration of the EV-AV into a connected and shared individual car system called CyCabs. Its initiator, Michel Parent, remembered that he was inspired by the JohnnyCab robot-taxi in the movie *Total Recall* (1990). Here too, liability concerns prevented further development, but Google's prototype X of 2014 "resembled the CyCab in many ways," and was promised to appear on the streets of California by 2020.[21] That was a decade after an AV's 'sensing body' performed successfully during a 'DARPA Grand Challenge' in the American Mojave desert. Kröger showed how "the techno-optimism of the last decades was only possible with a lack of historical knowledge." A nice example of this is the assumption that the autonomous EV will more or less automatically solve the fossil paradigm problem, just as the fossil car solved the urban horse pollution problem at the end of the nineteenth century.[22] And although 'level 5,' the highest scale of 'full automation' in the taxonomy of the Society of Automotive Engineers (SAE), is "probably a myth," according to Google's Waymo CEO in 2019, connected AV proponents promise the elimination of fatal accidents, the inclusion of new types of users (such as the elderly, or the physically impaired), and a 'just' cohabitation with other transport modes, especially cyclists and pedestrians.[23] Critics oppose the tendency to solve the climate crisis by 'technical fixes.'[24]

Again, a historical approach from a global point of view would have prevented researchers and AV proponents from neglecting an entire history of jitneys, shared rides, and other forms of (socially) 'connected' forms of mobility, in the West, but especially in the Global South.

But user culture can be very resilient, especially the historically deeply rooted 'adventurousness' of the gasoline car, nowadays investigated under the banner of "petro-masculinity" and "combustion masculinity." Together with the AV extension, observers believe that EVs nonetheless will bring about a "'demasculinization' of automobility," a belief criticized by those who see in a cultural reluctance to adapt a much more serious barrier to embracing the "femininity" of AV-EVs, a feeling that is most strong among "men who do not consider purchasing EVs" (an echo of Virgina Scharff's alibi argument, see Chapter 1).[25] Just like the optimism aroused by the sudden decline of car acquisition among Western youth around the turn of the century (in the literature known as "peak car"), the resilience of automobilism is much stronger than often assumed.[26] Feminists have underscored this resilience even more, by observing how "the science and engineering focus has dominated climate discourse" and by "explicitly linking optimism in technology with masculinity and a patriarchal, controlling way of thinking."[27]

With the electric vehicle in shared use we are already firmly moving into a possible future. After all, *expectations* form a crucial element of innovative behavior in the history of technology, especially among engineers and other technically inclined researchers, as they are working on technologies they have to treat as if they are already realized.[28] It would disrupt the limits of this chapter to attempt a comprehensive history of automotive futures (which should include the blossoming of scholarly futurism in the West [à la Alvin Toffler] or the Marxist *prognostik* in the East, areas of study John Urry proposed should "be reclaimed for social science and, in a way, for people in their day-to-day lives," as well as the later sociology of expectation),[29] but some common characteristics are worth mentioning for a full understanding of what follows.

To begin by staying with the industry itself: there exists a long history of 'concept cars,' many of them explicitly conceived as a test of a possible future, to be exercised during car shows among the public, varying from cars with atomic propulsion during the 1950s to personalized miniature city cars half a century later. Individual designers also have a reputation for claiming to "reinvent the automobile."[30]

At some more distance to the industry, researchers and research institutes published and publish their predictions, from the futurists and related publicists (such as the Club of Rome with their *Limits to Growth* in 1972, see Chapter 4) to Transition Studies scholars attempting to force open the future through strategic policy advice about 'niche marketing.' Within the realm of New Mobility Studies such prediction efforts start to be proposed as well, especially within the 'mobility justice' subfield, from "utopias" of "collective and socialized automobility" via a plea for "a shift to integrated transport and away from privately owned vehicles" to a critique of the concept of sustainability as a way of extending capitalist automobility.[31] In the following we will deal with five examples of such attempts at projections into the future, all of them reaching beyond the present 'car system,' so beyond the spectrum of alternatives offered by mainstream transport engineering and related perspectives, such as preferences for technical fixes à la Bill Gates. In presenting these perspectives I am especially interested in how far they are embedded in historical analysis.

Although the claim of the undesirability of automobilism is as old as automobilism itself, and was intensified during the 1960s and 1970s by global critics such as André Gorz and Ivan Illich (see Chapter 4), the tone for the current *fundamental* car critique has been set by Steffen Böhm, Campbell Jones, Chris Land, and Mat Paterson's "Impossibilities of Automobility" introduction to a collection of essays called *Against Automobility* from 2006.[32] The collection, they claimed, intended to theorize ("there are automobilities that do not depend upon the car") and politicize 'automobility' (through Foucault's theory about "truth, power and subjectivity"). They argued that the car "regime" is "inherently fragile" and thus "ultimately impossible *in its own terms*," because of four "antagonisms": congestion, sustainability, unsafety as a "killing machine," and geopolitical instability because of its oil dependence.[33] Formulated thus, such glances into the future are fatally unhistorical: these antagonisms existed for many decades and never really endangered the seemingly *inherently stable* car regime of the last hundred years.

In the following, I will deal with the five projections into the future as elaborations of, and answers to, the historically elaborated themes in the previous pages.

The first projection is the scholarly and activist promotion of the bicycle and other forms of 'micromobility' (small and lightweight vehicles not driving faster than, say, 25 km/h) as an alternative to the car, especially in an urban context. Typically, in an evolutionary scenario, electric bicycles are proposed to replace the car for more than half of its trips (those that are less than 8 km), offering "unique pleasure to their users, such as riding in the open air, control of the vehicle, liberty, and flexibility during their travels."[34] E-scooters promise to replace longer (commuter) car trips, but they also appear to replace trips on foot, on the bicycle, and by public transport.[35] Such proposals are often connected with the concept of 'smart cities,' including 'intelligent urban technologies,' which aim at "overcoming the need for car-based travel for fragmented work activities."[36] They fit in a general tendency to go beyond mainstream transport planning paradigms ("almost entirely a male domain") based on the economics of 'travel costs' and the focus on time gains, resulting in a fetishism of 'speed.'[37]

No less typically, in a more revolutionary scenario, cities (announce to) ban cars in large areas, thus encouraging and enforcing 'micromobilities' and pedestrianism, a policy that in a way is an extension of already much longer existing policies to create 'pedestrians zones.' Some architects and urban planners connect this with the concept of 'shared space' and even a '15-minute city' through a 'proximity revolution.' Although the proliferation of such policies in the West is presented as revolutionary, research in this direction may fall into the trap of what I proposed to call 'mobility Orientalism,' as it neglects a long tradition of traffic practices in the Global South, as we saw, thus perpetuating "that assessments of benefits and risks, R&D roadmaps, and regulatory measures [take place] with a 'global planner' or EU and US focused perspective that functionally prioritizes sectors, governments, and publics in the industrialized countries of the Global North."[38] Such alternatives to the car *system* rather than the car itself are meant to counter "the declining state of streets and streetscapes," caused (it is implied) by traffic engineers who have dominated urban design to the detriment of urban planners.[39] Although 'planners' kept working in the shadow of the 'builders' (the transport engineers; the road construction engineers) they slowly managed to regain power at the municipal level, where they proposed measures sometimes interpreted as part of a guerilla war against the "technicians." 'Traffic calming' through special meandering designs of streets was one of these measures, which made two Dutch planners hardly known in the Netherlands, Joost Vahl and Jan Giskes, world-famous: "Designing livable streets should never be left solely to traffic engineers."[40]

Historically, micromobility and the promotion of walking and cycling are the result of a growing self-confidence of planners in their struggle against the technicians of flow, including the "ecological version[s] of neo-liberalism."[41] Such shifts in the urban planning paradigm can be seen as an effort to counter the ongoing decay of cities, resulting in replacing parking lots with social housing and even the removal of urban highways, followed by compensation initiatives through providing affordable housing.[42]

The second example is a fourfold projection into future urban mobilities, all four accompanied by a currently existing city, a special form of mobility, and a characteristic literary or filmic utterance. Like the previous example it is equally evolutionary, as it hopes to gradually push the car from the street. Sociologist and cofounder of the 'mobility turn' John Urry, invited to write a mobility forecast by the British government, warned that the "new system," introduced "after the car," in order to be realistic "should be fashionable and faddish, and win hearts and minds by being more fun, which was the case with smartphones as opposed to landline phones."[43] This is certainly not the case with his fourth default scenario, Fortress City, predicted to occur if the other three are not realized. It is a future in which the current inequalities (represented in this case by the mobility segregation of Sao Paulo or the devastation of Detroit) are extrapolated into a world à la *Mad Max 2*, a 'neo-Mediaevalist' vision of warfare over shortages of energy and fuel. Urry did not sketch a mobility style for this scenario, but one can guess it might be one of the 'triumphalist' mobilities with drones and armored SUVs and Humvees, as in America's war in Iraq described in Chapter 4, or, more in general, the carnivalesque extremes of Western contemporary mobility. In the Fast-Mobility City (represented by the 'vertical city' of Dubai and described in J. G. Ballard's dystopian novel *High-Rise* [1975]) automobilism still has a high status, despite the ubiquitous congestion and despite the EVs running on the hydrogen-fed fuel cell. In the Digital City physical movement has been largely replaced by digital communication, an extrapolation of the electrified 'smart city,' represented by New Zealand's Christchurch as rebuilt after the earthquakes of 2010 and 2011. Smartphones now give status, an extrapolation of what we in Chapter 4 described as the brief (but according to Urry everlasting: "the car is no longer a 'must-have' status symbol") decrease of car ownership and car driving among the Western youth, but perhaps also the tendency of the 'absent-present car' that we observed in America's middle-class literature and songs.[44] We may add Gary Shteyngart's *Super Sad True Love Story* (2010) or Thomas Pynchon's *Bleeding Edge* (2013) as literary guides into such a future. Last, the Livable City seems to be Urry's preference (although he posits that "[n]one of these futures is simply preferable or probable"[45]): it is "partly or wholly free of 'cars,'" characterized by "'post-suburban' social lives," in its small-scale neighborhoods reminiscent of Jane Jacobs's recommendations (see Chapter 4) and represented by parts of New York, Bogotá, or Curitiba, or the 'shared space' in Vancouver. Remarkably, Urry does not mention a literary or filmic representation for this fourth scenario; instead, the works by Mimi Sheller and some Transition Studies scholars, urbanist Charles Montgomery's *Happy City* (2013), and former US Vice President Al Gore are mentioned.[46] Like these commentators, Urry thinks as a car driver.[47]

The third example of a future projection is a social-psychological study of the current condition of automobilism that ends with the end-of-the-car perspective à la Urry's Livable City: geographer and biologist Stefan Gössling's *The Psychology of the Car* from 2017.[48] At first sight, the study seems to mobilize the same psychosocial arguments that govern this book: the 'attachment' to (and on occasion: the 'addiction' to) 'automobility' make radical change very difficult if not impossible. But the study's 'attachment,' so to speak, to the 'mobility paradigm' also seems to include a reluctance to deal with the *history* of automobilism and thus misses the changes and the splits that have taken place in national and transnational car cultures in the course of the twentieth century, including the class base of this culture (as a basically middle-class culture, only recently split into a multiclass phenomenon). Gössling is certainly not alone in this: the same applies to his predecessors, from Peter Marsh and Peter Collett's popular *Driving Passion: The Psychology of the Car* (1986) to John C. Burnham's scholarly *Accident Prone: A History of Technology, Psychology, and the Misfits of the Machine Age* (2009).

From the psychologist's perspective, road rage and expressions of "anger" are conceived as a "personality trait." The study reproduces the industry's power through the questionable assumption that it is the 'automobility system,' more particularly "the infrastructure change based on politics supporting automobility [that] was decisive for the expansion of the automotive system," which delegates car culture, and the agency of the driver-consumer as car-and-driver, to a secondary role.[49] By making a distinction between 'attachment' and 'addiction,' and suggesting that the former is a psychologist's term for 'the love of the automobile,' while the latter is a dysfunctional aberration, Gössling places himself in the tradition of the early safety experts who for a long time (and continued by 'common sense' until the present) have separated a special category of so-called 'accident-prone' drivers from the majority of drivers, thus in fact defending and reproducing the 'automotive system' as such. Given his reluctance to restrict car use because of motorists' "affective and symbolic values associated with owned vehicles," Gössling is certainly not the first scholar to distinguish between "real" and "perceived" personal car dependence, the former a function of "basic life needs" (resonating with all those who see the car evolve from a plaything to a 'necessity'), the latter the preferred target for the discouragement or prohibition of car use. Nowhere is it made clear, however, how this reluctance to change car culture (a "taboo" for politicians) would enable the radical solution Gössling proposes: "the challenge is not make the current transport system more sustainable but rather to design a system that is altogether different." Against "the dominating principle of car ownership" stands "the importance, in cities, . . . to establish bicycle cultures."[50]

Our last and most radical example of a future projection is Robert Braun and Richard Randell's 'philosophical project' to "deactivate" the car system.[51] Although critical of the 'mobility paradigm,' their study is one of the few in this tradition that seems to reject the universalizing of current mobility by placing this paradigm squarely in "late-modernity." Before the founding publications of the 'mobility turn' by Sheller and Urry, they claim, "there was no automobility."[52] This seems to run counter to the trend among philosophers to support the current universalization of (Western) 'mobility.'[53] However, their method to reach this goal is not a history as we have defined and used it in this book, but a rather linear and finalist Foucaultian "alternative genealogy," as they call it, without much contingency (the many paths not taken) or attention to historical change (instead, they see history as a concatenation of 'events,' or in their words: "a succession of sociotechnical imaginaries"), without an acknowledgment of the diverging and converging interactions between different car (sub)cultures, thus falling into the diffusionist trap. Also, like most of the studies quoted here, they mostly rely on American sources, neglecting quite different European histories, let alone the diverging story of an entire Global Southern 'periphery.' In short: they intended to deliver "a presentist reading of past texts," thus constructing a "pre-history" of an "automobility *avant la lettre*."[54] This neglect of the historian's craft is a pity because their analysis is one of the most eloquent contributions to the debate about what automobility (both as a 'real' system and as an "imaginary," the latter considered to be an indistinguishable part of the former) nowadays *is*. Its 'pre-history,' they claim (following Paul Virilio), is one of speed, power, and violence, culminating in the Futurist and Fascist "dromocracy" (a society governed by "the violence of speed"). They see this society morph (way beyond Virilio) into a 'normalized' totalitarian state of exception, conceived as an Agambenian "concentration camp" and a "police state," which has meanwhile become a "global, singular" phenomenon.[55] How Marinetti's 'fascism' could become mainstream is not made clear, which one should expect from a history that goes way beyond other historians' claims (this historian among them) that Marinetti's is only one form (if not a cliché) of automobilism, preceded by other artists in Belgium and France who also practiced such a form of extremely violent automobilism, but were not related to Fascism, and instead were considered 'normal' or even 'red' novelists (see Chapter 1). Nor is the (very insightful, based on the sociologist Carl Schmitt) analysis of the freeway as a 'space of exception' new: it has been invoked as drivers' experience very convincingly by novelists such as Joan Didion. Declaring 'automobility' a planetary "form-of-life" globalizes and absolutizes a middle-class car experience, indeed, in the tradition of Agamben's observation of a global 'petty bourgeoisie' (see Chapter 4), and is a form of mobility Orientalism.[56]

There is no way to learn from this study how this "new, hegemonic, violent and totalitarian democratic order spread throughout cities, rural spaces and across the entire planet," without any crucial modifications. This cannot surprise the attentive reader of a study that starts its analysis in ancient Greece and not, say, in Confucian China or Buddhist India. Yet the metaphor of a 'spread over the planet' might have been inspired by the more recent 'epidemiological turn' among global road safety experts working in or for the WHO and the UN (see Chapter 4) and the redefinition of road unsafety as a public health problem.[57] Indeed, depicting a redefined 'automobility' as a holocaust is based on a very particular and very unhistorical reading of the road safety discourse and its recent history. Like Virilio and many more commentators comparing the 'onslaught' on the road with war (quoting Giorgio Agamben), the study cites (mostly American) experts who claim that nearly all road accidents are 'caused' by 'human error.'[58] Many of the cited experts appear to come from the field of 'industrial safety modeling,' entering the automotive field of road safety research as late as the 1990s when the initial measures of fatalities reduction since the early 1970s in the West (such as enforced helmet use on motorcycles, blood alcohol restrictions, and speed limits proposed by the first generation of postwar road safety experts) had lost much of their effect and more sophisticated, fine-tuned methods were called for, ushering in complex programs to promise an elimination of traffic deaths altogether (Scandinavian Vision Zero) or to a great extent (Dutch Sustainable Safety), as we saw in Chapter 4. Whereas the first and second generations of safety experts who had proposed and helped implement these programs were never able to fully explain the reduction of deaths and injuries after the 1970s, other researchers from the field of industrial safety science were able to develop sophisticated methods that also proved successful in aviation, and which they then started to apply to road mobility. They were supported by governments and other stakeholders (such as the car industry) who could use, I venture to claim, this alleged continuation and enormous expansion of the prewar 'accident-prone' myth to nearly the entire motoring population as an argument in favor of driverless and autonomous cars.[59] Yet, in entering the field of road safety research, these researchers had to adjust their approach to the consensus among the traditional road safety experts, who meanwhile had abandoned the 'blaming the victim' approach and had developed a systems approach, in an effort to design a 'foolproof' traffic system, just as other manufacturers had done with microwave ovens and other 'dangerous' household machines. Thus, 'causation modelers' nowadays use a "systems approach [as] the dominant approach in most safety critical domains" although "in practice, the persons approach [focusing on the individual human error] is still dominant in many areas, for example health care . . . and road transport."[60] Why this is so is unclear at the current state of historical research, but as I already remarked, there may be a connection with the industry's and related authorities' interest in driverless cars. But by far the majority of established road safety expertise, as discussed in *Accident Analysis & Prevention*, the *Journal of Safety Research*, *IATSS Research* (IATSS stands for International Association of Traffic and Safety Sciences, founded by the Japanese car industry), *Traffic Injury Prevention*, and *Zeitschrift für Verkehrssicherheit* as the most cited engineering journals, follows a very pragmatic, goal-oriented approach, with the Northwestern European countries celebrated within road-safety-related EU, WHO, and UN circles as the 'champions' of road safety policy, a self-sufficiency only lately questioned by an emerging expert community in the Global South, as we saw in Chapter 4. The latter are sometimes more outspoken proponents of a 'shared space' approach, which lately has been celebrated as a Western invention, but in historical reality is a reflection of a 'Southern' form of mixed traffic in which the car does not dominate.[61]

Braun and Randell not only misinterpret this history,[62] they also ignore or neglect a long tradition of historical STS research, in which the co-constructing role of an 'active user' and 'consumer' is stressed, not so much to blame them for being responsible for 'systemic errors' (as the 'causation' researchers and their followers seem to do when clinging to their 'person approach'), but to show how they have constantly co-constructed technology (including automotive technology).[63] Depicting them as victims of a safety holocaust and constructing a dichotomy between the "docile bodies that are repressed" of pedestrians and cyclists on the one hand and the 'witnesses and collaborators' of automotive fascism on the other (realizing that at least in the West these roles are 'situational': many motorists are also cyclists and pedestrians, and vice versa) would only strengthen the epistemic and political power of the manufacturers and related institutional stakeholders, such as "the engineer, the road safety expert, the corporation, the advertiser and the state."[64] The emphasis on the systemic aspects of automobilism tends to obscure and make invisible the human actors in this history.

Compared to Urry's four scenarios, Braun and Randell's study is particularly weak when it comes to the question of how to reach 'post-automobility.' They concede that they cannot offer "a new imaginary . . . or a more beautiful dreamscape," and recommend activism either based on 'happenings'-type actions à la the Amsterdam Provo movement of the 1960s ("put a vase of roses in a spot where cars usually park") or Bakhtinian-carnivalesque "metropolitan rebellion" in order to "disrupt the historical order," taking inspiration from Michael Hardt and Antonio Negri's and Anna Nikolaeva's "commoning" and the recent revival of urban 'shared space.' After the finishing of their book, they added the suggestion of a form of activism borrowed from the anti-tobacco campaign, including the placing of health warnings on cars and other actions to counteract the spread of misinformation and doubt about the true effects of automobilism, stemming from car industry supporters. That is why, they claim, the promises from the industry about 'Vision Zero' and zero emission through EV promotion should not be believed: such claims "fail even on their own terms." Such distrust toward the industry and related "powerful interests" is further enhanced because of the "diversionary reframing" and the "outright manipulation of information" by "climate deniers." The future is not, Braun and Randell emphasize, "a world of automated vehicles but one where humans re-acquire the political to act freely in concert," for instance by "[b]iking in the middle of the road slowing down traffic; setting up temporary gardens on parking spots during the weekend; driving at slow speeds—not breaking the rules but also not conforming with the coercive practices that determine the realities of the road; signaling the space around the bicycle to those who cross over the imaginary line in their cars."[65]

This is not how one imagines to 'deactivate automobility' in order to save the planet before the critical years of 2030/35. I don't blame them; I myself don't know a faster way either. As a historian, putting on a 'normative' hat, I have commented on the fatal decision around the turn of the twentieth century to go for the gasoline car, whereas engineers and other experts saw the electric version as the best option, convinced that the gas car was only a fad of the rich, just like the bicycle, and would soon disappear. How right they were, and how wrong. Likewise, the freeway as a cocoon for high-speed driving was first rejected on political grounds as a fascist project by the road engineers within the transnational PIARC, but was then accepted as an effective tool to increase road safety. How right they were, and how wrong: the car could much more easily have been abandoned if the world had not been paved with an incredibly costly railway-like asphalt-concrete network. Indeed, I do not have a 'quick fix' either, although I wonder why engineers and planners are excluded from the move away from the car and put aside as "engineering elites." Proposed by a 'philosophizing elite.'[66] "[W]e must also acknowledge," Laura Lindegaard warned, "the more ephemeral and illusive, everyday discursive accomplishments of rationalities in the automobility regime in order to ever end the regime."[67]

Deactivating mobility was recently achieved by the COVID-19 virus and by national governments in its name. Ironically, these same governments later recommended using the car (and not public transport) as a protective capsule against contagion.[68] Meanwhile, after the pandemic, automobilism was back to 'normal,' and increasing, worldwide.

Jason Henderson has shown how in the case of Atlanta, Georgia (the metropolis perhaps the most car-dependent in the US, punished by withholding federal road funds because of its 'excess' of automobilism), "many scholars, planners and public officials . . . are frustrated by the claim of a universal car culture making political challenges to automobility seem futile." He analyzed how behind the "inevitability hypothesis" in the discourse about automobilism is hidden (what we have come to call in the current study) a racialized 'ironic car,' a form of "secessionist automobility" that protects white middle-class drivers from the racialized 'Other.'[69] And yet, as Goodwin has emphasized, it is in the cities where the struggle against the car could start, and in fact already has started as we saw earlier in this chapter. There, as she observed (against Urry's path dependency thesis), people can and do undertake actions that have an immediate, visible, and livable effect on people.[70] The 'global,' anthropologist Noel Salazar posits, is too big as a place of action: "it is hard and dangerous to talk about mobilities and what mobilities mean and how they function from a global perspective because there are huge differences."[71] The German sociologist Katharina Manderscheid bases her historically inspired analysis of 'the movement problem' (in which she also mobilized several German contributors to the debate, not often found in the hegemonic Anglophone discourse) on Goodwin to propose a "post-consumerist future" that, per Goodwin, should start in the cities, where "[u]nequal access to a car," contrary to an all too universalized 'justice' approach, "does not by itself imply unequal social chances." Her (admittedly "rather Western and European") thesis that Foucaultian "governmentality," which resonated with Cotten Seiler's historical analysis, should be "grounded in thorough historical research," resulting, for instance, in "acknowledging that the situation looks different elsewhere," could provide a solid basis for transformative activism.[72] I cannot agree more.

As a historian, who has 'witnessed' more failed than successful projections into the future, I am reluctant to take part in the scenario bugaboo, all the more so because presenting a couple of scenarios will prove one always right. Instead, I have collected the most salient points of the previously sketched world history of automobilism, convinced that ignoring them would jeopardize us being equipped properly for the mobility future. Two of these 'salient points' are primordial. First: the insight, stemming from a historical analysis of the West and later also of automotive globalization, that the middle class not only was the 'inventor' of both the cycling and the automotive movements, but also the crucial agency for the spread of automobilism over the globe. It is onto the middle-class car culture that any effort to change the car paradigm should hook. It is there that the 'adventure machine' was conceived, and there that its 'taming' was set in motion, resulting in a covert adventurous substrate hidden by the ubiquitous, absent-present car. This class will do anything to perpetuate its culture, including electrifying its adventure machine and all other measures to redefine this adventure along sustainable trajectories, from changing to smaller cars, using (human-powered or electric) bicycles in the city, and even (human-controlled or autonomous) car sharing.

The second salient point is 'informal' transport, so widespread in the ever growing cities in the developing world (and also, to a certain extent, in Western cities) that its name is a caricature of its actual use. This form of mobility is related to rural–urban migration, forms one of the few opportunities of acquiring a livelihood for the urban poor, and necessitates another scholarly approach than a focus on 'adventure,' in the tradition of labor history. The Western automotive bias ignores the insight, taken from a historical analysis of the last half century, that the millions of riders and drivers of motorcycle taxis, jeepneys, *matatus*, and *fula fula* will never give up their two-stroke mopeds, their racing 'pirate taxis,' and their auto-rickshaws without a fight, despite some efforts from below to electrify their vehicles.

In order to attack such and many other salient points and to equip itself with the tools to work toward a more benign and just future, mobility history should be ready for a change, in several directions.

The first directional change is toward a more inclusive approach to mobilities, in the tradition of subaltern studies, such as historical analyses of indigenous mobilities.[73] This book contains many examples of 'reading against the grain' in order to go beyond mobility Orientalism. The scholars who analyze and criticize these developments belong to the same (middle) class, which should make any critical observer suspicious.

The second directional change is a closer cooperation with the social sciences, currently in the full flow of change themselves. One of these changes is the development of so-called 'mobile methods' among ethnographers, allowing a much more detailed opening toward "what seems to really take place" in people's daily mobilities, without giving up the more conventional methods, as Peter Merriman has cautioned.[74] Such studies also bring new theoretical insights, such as that "the vehicles themselves, as well as the transit space they operate from" in the case of transport hubs in Africa, "are in effect 'meeting points' for daily conversations."[75] Studying them should be part of "decolonizing" mobility knowledge, which makes "leadership by transport researchers from/in territories under colonialism or coloniality . . . highly desirable."[76] Getting rid of a Western car bias would enable the design of studies "with particular focus on the wellbeing of the poor," rather than the relationship between car use and obesity.[77]

The third directional change is to work around the clearly existing fear of 'technical fixes' among social scientists and historians alike by including transport experts and urban planners in the orbit of New Mobility Studies. A nice example is Tanu Priya Uteng and Karen Lucas's recent *Urban Mobilities in the Global South*, "making a jump . . . to studying the impacts of the nexus between infrastructure and empowerment."[78] Although early planning, especially in a colonial context, was racist and segregationist, urban planners lately seem to be regaining some self-confidence after they lost the battle over urban design in the 1920s.[79]

Recently, ethnographic research and 'mobile methods' (for instance on foot, in postsocialist Kosovo)[80] have provided us with a much richer impression of what mobility is all about. Such research resonates with Francis Nyamnjoh's observation that "the metaphor of flow," so ubiquitous in current research on mobilities, "ignores the reality of global inequalities, marginalization and disconnection, and the fact of an uneven playing field that make the process more one of global hopping than global flows."[81] Such research, resulting in the emergence of a new genre of 'social science travelogues,' also requires narrative, near-literary skills, such as in Jeremy Morris's accounts of Russian everyday postsocialism.[82] We need many more of such studies, enriched by historical sensitivity.

Notes

1. Katherine J. Goodwin, "Reconstructing Automobility: The Making and Breaking of Modern Transportation," *Global Environmental Politics* 10, no. 4 (November 2010): 60–78, here 60.
2. See for the following also: Gijs Mom, "New Mobility Studies: Niche to Mainstream; History and Social Sciences in the Mobility Turn" (*Transfers*, forthcoming).
3. For the hype character of the mobility topic (representing 11 percent of all topic clusters) in the preferred journal by transition students see Ivan Savin and Jeroen van den Berg, "Main Topics in EIST during Its First Decade: A Computational-Linguistic Analysis," *Environmental Innovation and Societal Transitions* 41 (2021): 10–17, here 12 (Figure 2); for an example of a history of recent EV development in which visions of the car industry and its related experts are echoed see Renato J. Orsato, Marc Dijk, René Kemp, and Masaru Yarime, "The Electrification of Automobility: The Bumpy Ride of Electric Vehicles toward Regime Transition," in Frank W. Geels, René Kemp, Geoff Dudley, and Glenn Lyons (eds.), *Automobility in Transition? A Socio-Technical Analysis of Sustainable Transport* (New York: Routledge, 2012), 205–28, here 209 (lead-acid batteries with their "very limited lifetime and range").
4. Johan Schot, "Confronting the Second Deep Transition through the Historical Imagination," *Technology and Culture* 57, no. 2 (April 2016): 445–56, here 449.
5. Frank W. Geels, "Processes and Patterns in Transitions and System Innovations: Refining the Co-Evolutionary Multi-Level Perspective," *Technological Forecasting and Social Change* 72 (2005): 681–96; Frank W. Geels, "The Multi-Level Perspective on Sustainable Transitions: Responses to Seven Criticisms," *Environmental Innovation and Societal Transitions* 1 (2011): 24–40; "poor conduct of historical case studies": Audley Genus and Anne-Marie Coles, "Rethinking the Multi-Level Perspective of Technological Transitions," *Research Policy* 37 (2008): 1436–45, here 1444.
6. Frank W. Geels, "A Socio-Technical Analysis of Low-Carbon Transitions: Introducing the Multi-Level Perspective into Transport Studies," *Journal of Transport Geography* 24 (2012): 471–82, here 471.
7. Cristina Temenos et al., "Theorizing Mobility Transitions: An Interdisciplinary Conversation," *Transfers* 7, no. 1 (Spring 2017): 113–29; see especially the comments by Tim Schwanen (116–17: "mistakes") and Tim Cresswell (118–19: "fixation on technology").
8. Gijs Mom, "De moderne elektro-auto," in Gijs Mom and Vincent van der Vinne, *De elektro-auto: Een paard van Troje?* (Deventer: Kluwer Voertuigtechniek, 1995), 9–110.
9. Timothy Mitchell, *Carbon Democracy: Political Power in the Age of Oil* (London: Verso, 2013 [2011]), 186.
10. Tjerk Gualthérie van Weezel, "Hoe 197 landen het eens werden over het einde van fossiele energie" [How 197 countries agreed on the end of fossil energy], *Volkskrant* (14 December 2023): 13–15; Mimi Sheller, *Advanced Introduction to Mobilities* (Cheltenham: Edward Elgar, 2021), 105.

11. Sandra Wappelhorst, "The End of the Road? An Overview of Combustion-Engine Car Phase-Out Announcements across Europe" (Briefing, International Council on Clean Transportation, ICCT, 10 May 2020) (https://theicct.org/publication/the-end-of-the-road-an-overview-of-combustion-engine-car-phase-out-announcements-across-europe/, last consulted 13 February 2025).
12. Camila Domonoske, "Giving up Gas-Powered Cars Was a Fringe Idea. It's Now on Its Way to Reality," *NPR*, 20 November 2021, https://www.npr.org/2021/11/20/1055718914/ (retrieved 20 March 2024).
13. Rebecca Heilweil, "It's the End of the Car as We Know It: The Electric Vehicle Transition is Really a U-Turn," *Vox*, 2 September 2022, https://www.vox.com/recode/23333356/ev-electric-vehicle-self-driving-cars-internal-combustion-driving (retrieved 20 March 2024); Bryan Appleyard, *The Car: The Rise and Fall of the Machine That Made the Modern World* (London: Weidenfeld & Nicolson, 2022), 5.
14. Brian T. Watson, *Headed into the Abyss: The Story of Our Time, and the Future We'll Face* (Swampscott, MA: Anvilside Press, 2019), 193 (Big Tech's "true motivation" to develop EV-AVs "is the money to be made").
15. Anika Regett, Wolfgang Mauch, and Ulrich Wagner, "Carbon Footprint of Electric Vehicles—A Plea for More Objectivity" (Forschungsstelle für Energiewirtschaft, 2019).
16. For a report from an EV lobby group (International Council on Clean Transportation, ICCT) see Georg Bieker, "A Global Comparison of the Life-Cycle Greenhouse Gas Emissions of Combustion Engine and Electric Passenger Cars" (Berlin: ICCT, 2021); for a report (based on only partial LCAs) from the automotive and oil industry, written by the Joint Research Centre (JRC) of the European Commission itself, see M. Prussi et al., "JEC Well-to-Wheels Report V5; Well-to-Wheels Analysis of Future Automotive Fuels and Powertrains in the European Context" (Luxemburg: Publications Office of the European Union, 2020), 9. I thank Auke Hoekstra and Carlo van de Weijer (both Eindhoven University of Technology) and Richard Smokers (TNO Delft) for their suggestions of reliable LCAs (emails on 3 and 4 March 2024). LCAs are "insufficient": Jorg Roosen, Wim Marneffe, and Lode Vereeck, "A Review of Comparative Vehicle Cost Analysis," *Transport Reviews* 35, no. 6 (2015): 720–48, here 732.
17. Jason Henderson, "EVs Are Not the Answer: A Mobility Justice Critique of Electric Vehicle Transitions," *Annals of the American Association of Geographers* 110, no. 6 (2020): 1993–2019, here 1993, 1994, 1997.
18. Ibid., 2003, 2005.
19. Ibid., 1995–97; new-car CO_2: Christian Brand, Jillian Anable, Ioanna Ketsopoulou, and Jim Watson, "Road to Zero or Road to Nowhere? Disrupting Transport and Energy in a Zero Carbon World," *Energy Policy* 139 (2020): 111334, 1–2, here 1; "corporate": Naveeda Kahn, *In Quest of a Shared Planet: Negotiating Climate from the Global South* (New York: Fordham University Press, 2023), 174.
20. Daniel Sperling et al., *Three Revolutions: Steering Automated, Shared, and Electric Vehicles to a Better Future* (Washington, DC: Island Press, 2018).
21. Fabian Kröger, *From Automated to Autonomous Driving: A Transnational Research History on Pioneers, Artifacts and Technological Change (1950–2000)* (Cham: Springer, 2023), 222; Daniel J. Fagnant and Kara Kockelman, "Preparing a Nation for Autonomous Vehicles: Opportunities, Barriers and Policy Recommendations," *Transportation Research Part A* 77 (2015): 167–81, here 168 (2020).

22. Fábio Duarte and Carlo Ratti, "The Impact of Autonomous Vehicles on Cities: A Review," *Journal of Urban Technology* 25, no. 4 (2018): 3–18, here 3, where an article by historians Joel Tarr and Clay McShane is cited. Typically, however, 'horse substitution' by the car was a process taking more than half a century, and the critique of the horse as a polluter only swelled after an alternative became available (before that moment, the manure problem formed an integral part of the 'horse economy'). This phenomenon also plays a role in the current situation (see later on in the text). Gijs Mom, "Compétition et coexistence: La motorisation des transports terrestres et le lent processus de substitution de la traction équine," *Mouvement Social* 229 (October/December 2009): 13–39; Kröger, *From Automated to Autonomous Driving*, 6 (lack of history).
23. Ibid., 10 ("myth").
24. Brand et al., "Road to Zero or Road to Nowhere?"
25. Cara Daggett, "Petro-Masculinity: Fossil Fuels and Authoritarian Desire," *Millennium: Journal of International Studies* 47, no. 1 (2018): 25–44; combustion: Matteo Vivi and Anne-Mette Hermans, "'Zero Emission, Zero Compromises': An Intersectional, Qualitative Exploration of Masculinities in Tesla's Consumer Stories," *Men and Masculinities* 25, no. 4 (2022): 622–44.
26. Peter Wells and Dimitrios Xenias, "From 'Freedom of the Open Road' to 'Cocooning': Understanding Resistance to Change in Personal Private Automobility," *Environmental Innovation and Societal Transitions* 16 (2015): 106–19.
27. Jennie C. Stephens, "The Dangers of Masculine Technological Optimism: Why Feminist, Antiracist Values Are Essential for Social Justice, Economic Justice, and Climate Justice," *Environmental Values* 33, no. 1 (2024): 58–70, here 62, 60.
28. Gijs Mom, "'The Future is a Shifting Panorama': The Role of Expectations in the History of Mobility," in Weert Canzler and Gert Schmidt (eds.), *Zukünfte des Automobils: Aussichten und Grenzen der autotechnischen Globalisierung* (Berlin: edition sigma, 2008), 31–58.
29. David C. Engerman, "Introduction: Histories of the Future and the Futures of History," *American Historical Review* 117 (December 2012): 1402–10, here 1403–4; Urry quoted by Michael Haldrup, "What Futures? Our Future, the Planet, and Cohabitation," *Transfers* 13, no. 1/2 (Spring/Summer 2023): 144–61, here 147; expectation: Johannes Kester, Benjamin K. Sovacool, Lance Noel, and Gerardo Zarazua de Rubens, "Between Hope, Hype, and Hell: Electric Mobility and the Interplay of Fear and Desire in Sustainable Transitions," *Environmental Innovation and Societal Transitions* 35 (2020): 88–102, here 88.
30. William J. Mitchell, Christopher E. Borroni-Bird, and Lawrence D. Burns, *Reinventing the Automobile: Personal Urban Mobility for the 21st Century* (Cambridge, MA: MIT Press, 2010).
31. Noel Cass and Katharina Manderscheid, "The Automobility System: Mobility Justice and Freedom under Sustainability," in Nancy Cook and David Butz (eds.), *Mobilities, Mobility Justice and Social Justice* (London: Routledge, 2019), 101–15, here 101, 107.
32. Steffen Böhm, Campbell Jones, Chris Land, and Mat Paterson, "Introduction: Impossibilities of Automobility," in Steffen Böhm, Campbell Jones, Chris Land, and Mat Paterson (eds.), *Against Automobility* (Malden, MA: Blackwell, 2006), 3–16.
33. Ibid., 4, 6, 9–11 (italics in original).
34. Buket Şengül and Hamid Mostofi, "Impacts of E-Micromobility on the Sustainability of Urban Transportation—A Systematic Review," *Applied Sciences* 11 (2021): 5851, 4.

35. Laura Gebhardt, Christian Wolf, and Robert Seiffert, "'I'll Take the E-Scooter Instead of My Car'—The Potential of E-Scooters as a Substitute for Car Trips in Germany," *Sustainability* 13 (2021): 7361, 5.
36. Tommi Inkinen, Tan Yigitcanlar, and Mark Wilson, "Smart Cities and Innovative Urban Technologies," *Journal of Urban Technology* 26, no. 2 (2019): 1–2, here 1.
37. Susan Hanson, "Transportation: Hooked on Speed, Eyeing Sustainability," in Eric Sheppard and Trevor K. J. Barnes (eds.), *A Companion to Economic Geography* (Oxford: Blackwell, 2000), 468–83, here 475.
38. Benjamin Sovacool, "Expanding Carbon Removal to the Global South: Thematic Concerns on Systems, Justice, and Climate Governance," *Energy and Climate Change* 4 (2023): 100103, 1–16, here 12.
39. Ben Hamilton-Baillie, "Shared Space: Reconciling People, Places and Traffic," *Built Environment* 34, no. 2 (2008): 161–81.
40. H. G. Vahl and J. Giskes, *Traffic Calming through Integrated Urban Planning* (Paris: Amarcande, April 1990 [1988]), 7, 10. The booklet is written as a guerilla handbook, with language such as "villages are destroyed in order to let motorized traffic pass through" (6) and "the bombardment of towns and villages by motor vehicles" (6), and advice to begin urban traffic reform in the neighborhood of schools (18).
41. Wojciech Kębłowski and David Bassens, "'All Transport Problems Are Essentially Mathematical': The Uneven Resonance of Academic Transport and Mobility Knowledge in Brussels," *Urban Geography* 39, no. 3 (2018): 413–37, here 422.
42. Steven Higashide, "Never Again Is Now: The Transportation Professions' Responsibility to Work Toward Justice," in Ryan Reft, Amanda K. Phillips de Lucas, and Rebecca C. Retzlaff (eds.), *Justice and the Interstates: The Racist Truth about Urban Highways* (Washington, DC: Island Press, 2023), 175–92, here 176, 181; decay: Vukan R. Vuchic, *Transportation for Livable Cities* (New Brunswick, NJ: Center for Urban Policy Research, Rutgers, State University of New Jersey, 1999), 3.
43. John Urry, *What is the Future?* (Cambridge: Polity Press, 2016), 136. See for the following ibid., 147–55. Also see Kingsley Dennis and John Urry, *After the Car* (Cambridge: Polity Press, 2009), Ch. 7 ("Scenarios," 131–64).
44. Urry, *What is the Future?* 144.
45. Ibid., 153.
46. There is some analogy with Giovanni Arrighi's economic history of the 'long twentieth century,' his narrative ending in three scenarios: "the formation of a world empire, the formation of a non-capitalist world economy; or a situation of endless systemic chaos." Giovanni Arrighi, *The Long Twentieth Century: Money, Power, and the Origins of Our Times* (London: Verso, 2010), 372. In a "Postscript" he opts for the second scenario to be the most likely (ibid., 385).
47. Janet Elspeth McAllister, "Stuck Fast: A Critical Analysis of the 'New Mobility Paradigm,'" master's thesis (Auckland: University of Auckland, 2011), 61, quoting Urry's "Social Networks, Travel and Talk" in *British Journal of Sociology* (2003), in which he argues that in a certain case travel ability is "all used up."
48. Stefan Gössling, *The Psychology of the Car: Automobile Admiration, Attachment, and Addiction* (Amsterdam: Elsevier, 2017).
49. Ibid., 19, 87.
50. Ibid., 229, 232, 240.
51. Robert Braun and Richard Randell, *Post-Automobility Futures: Technology, Power, and Imaginaries* (Lanham, MD: Rowman and Littlefield, 2022), 108, 146.

52. Ibid., 14.
53. See for instance very recently Thomas Nail, "What is the Philosophy of Movement?" *Mobility Humanities* 1, no. 1 (January 2022): 6–22: "we sense that the pace of life is speeding up" (7) and "in the philosophy of movement, motion is primary, indeterminate, and without any other cause or explanation. Neither time, duration or vitality is more primary" (15). "Reality is mobility itself," Bergson argued (quoted on 15). For a similar critique of the problematic universalizing of concepts ('society,' 'agency,' but also 'Orientalism,' and why not 'mobility?') in the social sciences see Charles Piot, *Remotely Global: Village Modernity in West Africa* (Chicago: University of Chicago Press, 1999), 2–7.
54. Braun and Randell, *Post-Automobility Futures*, 110 ("late-modernity"), 19 ("succession"), 34 ("genealogy"), 9 ("presentist," "pre-history"; italics in original).
55. Ibid., 93–94 ("camp"), 109 ("violence of speed"), 112 ("singular"), 116 ("police state").
56. Ibid., 76, 92–93.
57. Ibid., 107; on Greece as the origin of civilization see Gyan Prakash, "Writing Post-Orientalist Histories of the Third World: Perspectives from Indian Historiography," *Comparative Studies in Society and History* 32, no. 2 (April 1990): 383–408, here 389.
58. Braun and Randell, *Post-Automobility Futures*, 84, 87, 96.
59. Eleni Petridou and Maria Moustaki, "Human Factors in the Causation of Road Traffic Crashes," *European Journal of Epidemiology* 16, no. 9 (2000): 819–26, here 823: "The concept of accident proneness needs further consideration."
60. Paul Salmon et al., "Managing Error on the Open Road: The Contribution of Human Error Models and Methods," *Safety Science* 48 (2010): 1225–35, here 1226.
61. Hamilton-Baillie, "Shared Space." See Ruud Filarski and Gijs Mom, *Geschiedenis van de verkeersveiligheid in Nederland* [History of traffic safety in the Netherlands] (forthcoming 2025).
62. Most transport ministries in the world nowadays, upon advice from the original road safety expert community, follow the exact opposite approach to what Braun and Randell (*Post-Automobility Futures*, 97) claim: "In locating violence within the individual driver, not the special ordering, automobility violence is rationalized and rendered acceptable." They also miss the 'epidemiological turn' as a specific contribution of the Global South, through the WHO and UN, to global road safety expertise. See on this turn Iris Borowy, "'To Study "Accidents" as a Whole May Be Like Studying "Disease" as a Whole': The Epidemic of Road Traffic Injuries," in Jörg Vögele, Stefanie Knöll, and Thorsten Noack (eds.), *Epidemics and Pandemics in Historical Perspective* (Herbolzheim: Centaurus Press, 2015), 177–189.
63. See for instance Kathleen Franz, *Tinkering: Consumers Reinvent the Early Automobile* (Philadelphia: University of Pennsylvania Press, 2005). Also see Chapter 1 for the social construction of technology.
64. Braun and Randell, *Post-Automobility Futures*, 95 ("docile"), 111 ("state"). Braun and Randell's ambiguous approach to the motorist's societal role is perhaps a manifestation of this misinterpretation: on the one hand she is a "witness and even collaborator in the totalitarian automobility order," on the other hand she is a victim (ibid., 117, 121). Since the 1970s in the West there are more deaths among the motorists than pedestrians and cyclists; this is clearly not the case in the Global South: P. L. Jacobsen, "Safety in Numbers: More Walkers and Bicyclists, Safer Walking and Bicycling," *Injury Prevention* 9 (2003): 205–9, here 205.

65. Braun and Randell, *Post-Automobility Futures*, xxv ("fail"), 159 ("vase," "dreamscape"), 147, 153, and 155 ("rebellion" and carnival), 152 (commoning and long quote), 150 ("freely"); anti-tobacco: Robert Braun and Richard Randell, "Towards Post-Automobility: Destituting Automobility," *Applied Mobilities* 8, no. 3 (2023): 201–17, here 210; climate deniers: Aaron M. McCright and Riley E. Dunlap, "Defeating Kyoto: The Conservative Movement's Impact on U.S. Climate Change Policy," *Social Problems* 50, no. 3 (August 2003): 348–73, here 351.
66. Braun and Randell, *Post-Automobility Futures*, 138. See for the very understandable but nonetheless unwelcome tendency of a strand of engineering-skeptical thought Stephens, "The Dangers of Masculine Technological Optimism."
67. Laura Bing Lindegaard, "The Discursive Accomplishment of Rationalities in the Automobility Regime," in Marcel Endres, Katharina Manderscheid, and Christophe Mincke (eds.), *The Mobilities Paradigm: Discourses and Ideologies* (London: Routledge, 2016), 68–87, here 83.
68. Gijs Mom, "Trending *Transfers*: A Decade of New Mobility Studies through the Lens of Transmodality, Transnationalism, and Transdisciplinarity," *Transfers* 10, no. 1 (Spring 2020): 2–21.
69. Jason Henderson, "Secessionist Automobility: Racism, Anti-Urbanism, and the Politics of Automobility in Atlanta, Georgia," *International Journal of Urban and Regional Research* 30, no. 2 (June 2006): 293–307, here 294, 297 ("too much automobility").
70. Goodwin, "Reconstructing Automobility," 74, 76.
71. Noel B. Salazar and Taehee Kim, "An Interview with Noel B. Salazar," *Mobility Humanities* 3, no. 1 (January 2024): 158–68, here 159.
72. Katharina Manderscheid, "The Movement Problem, the Car and Future Mobility Regimes: Automobility as Dispositif and Mode of Regulation," *Mobilities* 9, no. 4 (2014): 604–26, here 607, 620, 621n8 ("grounded"), 621n11 (Seiler), 621n12 ("elsewhere"); Cotten Seiler, *Republic of Drivers: A Cultural History of Automobility in America* (Chicago: University of Chicago Press, 2008).
73. For instance Georgine Clarsen, "*Black As*: Performing Indigenous Difference," in Nancy Cook and David Butz (eds.), *Mobilities, Mobility Justice and Social Justice* (London: Routledge, 2019), 159–72.
74. David Butz and Nancy Cook, "Mobile Methods, Epistemic Justice and Mobility Justice," in Cook and Butz (eds.), *Mobilities, Mobility Justice and Social Justice*, 81–97; Peter Merriman, "Rethinking Mobile Methods," *Mobilities* 9, no. 3 (2013): 167–87; quote: AbdouMaliq Simone, "The Urbanity of Movement: Dynamic Frontiers in Contemporary Africa," *Journal of Planning Education and Research* 31, no. 4 (2011): 379–91, here 379.
75. Daniel E. Agbiboa, "Introduction: Transport, Transgression and Politics in African Cities," in Daniel E. Agbiboa (ed.), *Transport, Transgression and Politics in African Cities: The Rhythm of Chaos* (Abingdon: Routledge, 2019), 1–16, here 5.
76. Tim Schwanen, "Towards Decolonised Knowledge about Transport," *Palgrave Communications: Humanities/Social Sciences/Business* 4 (2018): 1–6, here 5.
77. Remy Sietchiping, Melissa Jane Permezel, and Claude Ngomsi, "Transport and Mobility in Sub-Saharan Cities: An Overview of Practices, Lessons and Options for Improvement," *Cities* 29 (2012): 183–89, here 183, 184.
78. For instance Jacqueline M. Klopp and Clemence M. Cavoli, "The Paratransit Puzzle: Mapping and Master Planning for Transportation in Maputo and Nairobi," in Tanu Priya Uteng and Karen Lucas (eds.), *Urban Mobilities in the Global South* (London: Routledge, 2017), 95–110; last quote: Maja Karoline Rynning, Tanu

Priya Uteng, and Karen Lucas, "Epilogue: Creating Planning Knowledge through Dialogues between Research and Practice," in ibid., 215–22, here 221. The traffic calming measures by Joost Vahl and others discussed in Chapter 4 were examples of "a power grab by urban planners and architects from traffic engineers. You don't leave the form of a building to the electrotechnical consultancy, don't you?" Tijs van den Boomen, "Op naar de leesbare straat" [Towards the readable street], *NRC Handelsblad* (2 June 2001) (my transation).

79. Daniel Oviedo, Caren Levy, and Julio D. Dávila, "Constructing Wellbeing, Deconstructing Urban (Im)Mobilities in Abuja, Nigeria," in Tanu Priya Uteng and Karen Lucas (eds.), *Urban Mobilities in the Global South* (London: Routledge, 2017), 173–94, here 174; Gijs Mom, "The Emancipation of the Urban View: Dutch Spatial Planning in an International Context (1920–1950)," in Jos Arts, Ruud Filarski, Hans Jeekel, and Bert Toussaint (eds.), *Builders and Planners: A History of Land-Use and Infrastructure Planning in the Netherlands* (Delft: Eburon, 2016), 85–149.
80. Rita Gagica and Ger Duijzings, "Liberated or Lawless? Social Life on Prishtina's Postwar Streets," in Ger Duijzings and Tauri Tuvikene (eds.), *If Cars Could Walk: Postsocialist Streets in Transformation* (New York: Berghahn, 2023), 46–69.
81. Francis B. Nyamnjoh, "Fiction and Reality of Mobility in Africa," *Citizenship Studies* 17, no. 6–7 (2013): 653–80, here 654.
82. Jeremy Morris, *Everyday Post-Socialism: Working-Class Communities in the Russian Margins* (London: Palgrave Macmillan/Springer Nature, 2016).

CONCLUSION

Past to Future, Atlantic to Pacific

As we have seen in the previous chapters, a world history of mobility modernization has to start with imperialism and migration. The former shaped colonial shipping and railways in Europe's and Japan's 'peripheries,' from the Atlantic Latin American and African countries to Pacific Asia. The latter saw the ships and trains used to travel to a better life elsewhere, from German and Irish workers to the United States to the Chinese poor looking for work in the fledgling urban transport sector in Singapore and Kolkata, Shanghai and Beijing, creating a collective culture of semilegal and semipublic rickshaw transport. These cultures dwindled and eroded when during the 1920s and 1930s an exogenous vehicle entered: the gasoline car, bought by the local elites.

Meanwhile, in the Western 'core' an equally semipublic urban transport culture emerged, based on electrically driven taxicabs, next to a fledgling bicycle culture driven by a new class in the full process of economic and political emancipation that soon started to 'upgrade' to motorized two- and four-wheel versions, remarkably quickly accommodated by their class fellows in the national governments. Within a few decades auto and touring clubs, an emerging car industry, and a World War co-constructed an *automotive system* enabling and fostering a very attractive (in the context of the fin de siècle and beyond) adventurous culture of temporal (high-speed), spatial (wandering in the countryside), and functional (tinkering) practices dominated by masculine, narcissistically aggressive, and violent fantasies of conquest, of women and land (peri-urban countryside, 'peripheral' colonies). We were able to reconstruct a quite nuanced picture of this culture through the analysis of novels and films. A parallel, more 'feminized' culture of gymkhanas and flower contests showed that a minority preferred a less aggressive, more 'domestic' behavior on the bicycle and on and in the car. Socialist cycle clubs developed a less aggressive cycling culture as well.

By the end of the 1930s the car system had evolved into a '*car society*,' after it had won the struggle of the systems, the so-called 'coordination crisis' of rail versus road. By then 'automobilism' had become 'automobility,' the term used since the 'mobility turn' of around 2000 by social scientists and historians alike. Such Western car societies not only were equipped with more-or-less countrywide highway systems enabling the lives of the car-driving upper and middle classes to be permeated by a car culture of holidaying, weekend spins, and an occasional commute; they were mostly white, and still predominantly male, as people of color were largely excluded from this culture and had to develop their own subculture, later called 'driving while black.' This car culture also permeated the lives of the working classes, through their participation in motor bus transport, their occasional participation in a motorcycle and moped culture, their labor as chauffeurs and drivers of trucks and buses, and not least the risk they encountered of being killed by cars when on foot or on the bicycle, men, women, and their children.

World War II was a war of motor vehicles (on land, in the air, and on and in the sea) and when most countries in the world were recovering from the belligerent devastations, it seemed at first that the ensuing decolonization gave the railroads and other forms of motorized public transport (buses, taxis) a new chance of becoming dominant, especially in socialist countries. In both the Soviet Union and China freight as well as passengers were predominantly transported by rail (and water), also in the cities, and even during the Cold War in the 'roadless' Soviet Union, after Stalin's death, there was still hope of a collective form of shared taxis and rental cars, spreading also to socialist countries in Eastern Europe. In socialist Tanzania a conscious anti-car policy was implemented. That neither the Soviets nor the CPC developed a socialist theory of mobility reflects the poverty of Marxism (which "has not really engaged with transport matters")[1] in this respect, although recent research by a new generation of mobility-oriented scholars unearthed the existence of a socialist urban planning tradition that favored the electric trolleybus and the diesel bus. Socialists and social democrats in the West either supported and protected public transport, or proposed an equal right for the working class to the possession of a family car (or they followed both strategies). Shared car use (and more so: shared bicycle use) became an option under capitalism for those who could not or would not own a car, including the failed 'dockless' experiments of Ofo and Mobike in China, which brought unheard of numbers of vehicles onto the streets.[2] It was an alternative of ownership, rather than use, although a flavor of the original 'socialist' culture remained, namely the availability of the car only in the case of need. Motorcycles, jitneys, and shared cars were (and are) intra-system alternatives, the proliferation of which was enforced (as economical 'demand') by the user.

But World War II had catapulted the United States into the position of hegemonic world power, and first its GIs, then the middle class and ever larger parts of the working class indulged in an automotive exuberance that modulated the 'car society' into an orgy of car mobility, enabled by very large and gas-guzzling cars, a very powerful oligopoly of industrial giants, part of a military-industrial complex aimed at world hegemony in a motor-vehicle-driven Cold War and hot wars in Korea and Vietnam, as well as a youth culture of high mobility supported by music and movies that functioned as effective propaganda over the world of a 'Western, prosperous way of life.' European countries followed suit, with smaller, more 'adventurous' cars and a more balanced mix of private and public transport.

When the car markets in the US and Europe threatened to become saturated, Western car propaganda was accompanied by massive efforts of 'development' of the rest of the globe, aimed at the construction of infrastructures including roadbuilding, sponsored by the World Bank, the International Monetary Fund, and Western consultants, somewhat later joined by the United Nations. The 'energy crises' of the 1970s hardly caused a dent in the national diffusion curves of car sales in the West, but they nonetheless formed a trigger for a very slow process of a rethinking of the car society, especially in the cities, increasingly pestered by congestion, pollution, and traffic deaths, even after most Western countries succeeded in diminishing the 'onslaughts' on the roads. This rethinking was primarily a middle-class affair (particularly through its ever more vocal intellectuals), and it was ambiguous, because while literature and movies showed a 'taming' of the 'adventure machine' and an invisibilization of the car as an 'absent car' (always there in the background), emancipatory movements among women, blacks and other people of color, Native Americans, and even children kept the machines of car production working. What then was advertised in movies and through other channels throughout the world was not so much this 'modest' automobilism, but the carnivalesque revival of hardcore adventure, mixed with a redefinition of the status traits of automobilism, created by cruisers, car modifiers, street racers, and all those other old and young 'deplorables' that inspired scenario writers to formulate the *Fast and Furious* utterances of Hollywood. The cruisers and modifiers revived the 'feminized' culture of early gymkhanas and flower contests.

It was this status seeking through car *ownership* (rather than *use*) that attracted local elites in the Global South to follow the example of their Western counterparts. They formed an enormous, ever growing reservoir of would-be motorists that fueled the precipitous roadbuilding and road motorization of Japan, South Korea, China, India, and many other Asian and Latin American countries, resulting in such an explosion of road traffic deaths that the World Health Organization and the United Nations reformulated this 'onslaught' into a public health problem, to be impugned just like cancer and obesity. After the crises of the 1970s, road network building in many of these countries led to serious debt and problems of road maintenance and a worsening of the urban crisis in both the Global South and the Global East, but with the rapid and spectacular emergence of car systems in China and related Asian countries a new geopolitical structure was in the making, driven by infrastructure proliferation and the restructuring of such states into so-called 'infrastructure states.'

And yet, it is much too early to speak of a 'planetary car society': next to the car, in many countries of the Global South a majority of other road users are counted, from pedestrians and cyclists to motorcyclists and vans and minibuses (and in some countries also still the rickshaw, now as the bicycle rickshaw or auto-rickshaw), the latter often functioning in a collective 'jitney' culture, giving a totally different traffic mix, which I called 'layeredness,' a mix that, on closer inspection, also exists in the West, albeit on a smaller scale, for instance in certain cities that are moving away from the car.

In the 'taming' of the Western 'adventure machine' it seems that the role of women and other historically marginalized groups such as ethnic minorities has been substantial, a shift from adventure to status (or perhaps better: a reemergence into the open of an always covertly present status now that the adventure has been tamed) that was continued when the car invaded the Global South. In order to understand this shift, and in general to understand the shift from an 'Atlantic' to a 'Pacific' culture of automobilism, analyses of literature and other belletristic utterances, as well as popular culture utterances such as songs, movies, and gaming, were crucial.

My somewhat polemical Chapter 5 hopefully convinced readers that solving the current planetary climate crisis and the role of automobilism therein can only be executed properly through an inclusion of mobility history in the scholarly and public debates that are necessary as a base of activism to reach collective and just forms of mobility, whether automotive or not. Solving this crisis should go through an acknowledgment of the crucial role of the middle classes, as well as the dominant existence of informal transport as a livelihood of the poor.

Other scholarly finds that could be 'harvested' through conceiving mobility history as a global one, based to a crucial extent on fictional sources, are the importance of labor history for the fruitful development of New Mobility Studies, as well as the constitutive role of 'mobility culture,' including the emergence of 'mobility Orientalism' and the need for a performative, normative character of mobility scholarship without falling into the trap of modal and other forms of partisanship. In fact the use of literature can be seen as a historical form of recently proposed 'mobile methods' in mobility research.

The history presented here leaned heavily on the subjectivist, agency-inspired approach, which indeed, *pace* Manderscheid, "may run the risk of neglecting underlying questions of the geopolitical order, political governance and socio-economic preconditions and their ongoing changes."[3] I believe that the past chapters contain enough geopolitical and other systemic arguments to counter this risk. Indeed, the recent COVID-19 pandemic showed us that radical change *can* be implemented 'from above,' by governments, supported by committees of doctors and engineers, but the rapid move back to normal after the pandemic suggests that the trigger to persistent change has to come from below, from subjective agency. Talking of agency: I know like no other what is lacking in the previous narrative, despite my 'readings against the grain' and my emphasis of subversive and subaltern mobilities. That is the voice of the 'other,' in the East as well as in the West. Not only her voice, but also her scholarship: solving the current mobility crisis has to leave behind the Western (and increasingly Southern and Eastern) car bias, and urgently needs cooperation between historians, social scientists, and transport experts and planners. It is the conviction of this author that the emphasis on automotive utilitarianism in the Global East and South (perhaps spiced by some of Urry's 'fun') should form the core of every plan for a post-automobility future.

Such plans should also involve a critical questioning of the field's 'received wisdom' that has largely been developed without the majority of earthlings in mind. Acknowledging that the majority walks may invite us to rethink the alleged universality of 'mobility': perhaps this majority has a preference to stay rather than move. Especially scholars from the Global South appear to see immobility rather than mobility dominating the globe.[4] Remarkably, mobility historians have hitherto not produced a serious analysis of changes in walking, as Barbara Schmucki observed already in 2012.[5] Nor have social scientists seriously reflected on the possibilities of worldwide violent resistance against measures of 'the elite' to make transport so expensive as to exclude the majority from its use, such as has happened with the bus price hikes in Latin America, the destruction of Rapid Transit buses in Africa, and the Yellow Vests in France.[6] From a world mobility history perspective, it is worthwhile to always keep in mind that the car is not dominant on the street, it is primarily dominant in our heads.

The New Mobility Studies field needs "more creative globalization research," to counter "the machine civilization that is the vehicle of [what Al-e Ahmad in 1962 called] Westoxication," so that we know what the rider of a motorcycle taxi in Lagos would think if he heard about a theatre production in the Netherlands, called *Zomeroffer* (Summer Sacrifice), in which in each show a car from someone from the audience is scrapped.[7]

Notes

1. Wojciech Kębłowski, Frédéric Dobruszkes, and Kobe Boussauw, "Moving Past Sustainable Transport Studies: Towards a Critical Perspective on Urban Transport," *Transportation Research Part A: Policy and Practice* 159 (2022): [reflowed pagination; 9].
2. Jun Zhang, "A History of Bicycle Mobility in Urban China: Shared Bikes, Technology, and Infrastructures (Part 2 of 2)" (Max Planck Institute for Social Anthropology, April 2022) (I thank Jun Zhang for sending me her manuscript).
3. Katharina Manderscheid, "The Movement Problem, the Car and Future Mobility Regimes: Automobility as Dispositif and Mode of Regulation," *Mobilities* 9, no. 4 (2014): 604–26, here 605.
4. Kerilyn Schewel, "Understanding Immobility: Moving beyond the Mobility Bias in Migration Studies," *International Migration Review* 54, no. 2 (2019): 328–55, here 333; see also Maximiliano E. Korstanje, *The Mobilities Paradox: A Critical Analysis* (Cheltenham: Edward Elgar, 2018), 7, 23.
5. Barbara Schmucki, "Against 'the Eviction of the Pedestrian'; The Pedestrians' Association and Walking Practices in Urban Britain after World War II," *Radical History Review* 114 (Fall 2012): 113–38, here 114.
6. Mimi Sheller, *Advanced Introduction to Mobilities* (Cheltenham: Edward Elgar, 2021), 117.
7. Raewyn Connell, *Southern Theory: The Global Dynamics of Knowledge in Social Science* (Cambridge: Polity Press, 2007), 216; Sander Janssens, "Kapot mooie auto" [Wrecking our lovely cars], *Volkskrant* (15 June 2024): V20.

Annotated Canon

Introducing Mobility History

Introduction

The publications presented here are meant to give the novice in *mobility history* a selection of reading material enabling her to gain a quick entrance into the field of (road) mobility history. Considered to be a subfield within New Mobility Studies (NMS) as defined by the first editorial of the journal *Transfers*, the set selected here (the majority in addition to the sources used for this book) does not cover adjacent fields such as migration studies or travel writing, and is very selective in other adjacent fields such as tourism. It is not a 'canon' (in the original meaning referring to a selection of primordial saints), nor is it based on a consensus within the field (such a consensus does not exist), although I believe that the publications given here are representative of the basic knowledge of most of the current experienced practitioners in the field. To this presumed 'common set' I added some publications that seem to have been forgotten and would merit, in my opinion, a rereading.

So, this is not a comprehensive bibliography: I deliberately chose only one or two examples per item (often the most comprehensive, sometimes accompanied by an iconic, pioneering example, apologizing to those colleagues that feel excluded), trusting that interested colleagues will find further references in the publication(s) I selected. For important items with a long tradition I selected one or two 'pioneers,' one or two that revived or redefined the subfield, and one or two recent sources for their comprehensive literature overviews.

What is largely left out is aviation, shipping, and to a large extent (although not fully) rail mobility (both urban and interurban). Nor does this overview provide an entrance into the general field of NMS (the 'mobility turn'), unless one understands 'entrance' as the way to approach NMS through a *historical understanding*. There are enough sources in which overviews of the social sciences and humanities within the 'new mobilities paradigm' can be found; this overview, on the contrary, has been constructed because of the apparent dearth of such overviews within the general NMS field (see the Introduction).

Many early sources we now experience as reductionist, finalist, and Whiggish, but novices to the subfield should nonetheless read them as bags of (historically evidenced) empirics, to avoid falling back on non-evidenced (no less Whiggish and finalist) 'popular' accounts.

Global, Transnational, and Multimodal Works

Apart from the trilogy on which this book is based (Mom, 2015, 2020, 2022, especially the latter two)[1] there does not exist a recent *world* mobility history. The only equivalent I can think of from the field of cycling studies is Smethurst's study (2015). In general, the series *Mobility in History* (published by Alphil for the first two years [2011 and 2012] and Berghahn Books for the remainder until 2015) provides an excellence entrance into the history of several areas, vehicle types, and topics.

At the beginning of post-WWII transport history some British economic historians published comprehensive *multimodal* histories, but they were always national in scope, for instance Dyos and Aldcroft, 1969, on British transport since the seventeenth century); or the edited volume by Theo Barker (1986), or Barker and Gerhold's overview (1993) of two centuries of road transport. Gerhold's edited volume on transport during the British horse-drawn era (1966) is another classic. Aldcroft and Freeman (1983) analyzed Britain's transport during the industrial revolution, a tradition revived in this century by Bagwell and Lyth (2002). Recently a French multimodal history of 'transports and mobilities' appeared by Baldasseroni et al. (2022). Another classic is Rose, Seely, and Barrett's book on America's transportation system (2006). A Dutch example is Filarski and Mom (2008, two volumes, the second as Mom and Filarski, in Dutch), on nineteenth-century inland water, rail, and road transport and twentieth-century rail and road transport (a summary in English by Filarski, 2011).

The same national approach goes for *railway* histories, but here some historians of technology have given us some early *transnational* histories from the perspective of imperialism (Adas, 1989; Headrick, 1981; Davis and Wilburn, 1991), recently added to with more up-to-date publications also stressing the transnationalism of the railways (Burri, Elsasser, and Gugerli, 2003, in German) and Japan's railway imperialism (McDonald, 2014). One of the Anglophone doyens of this strand of research was Colin Divall, professor at York's railway museum and university, who proposed a 'cultural turn' in 2005 (Divall and Revill).

Transnational road transport histories are at least as scarce, but in this century some *comparative studies* have been published, on the political economy of rail and road transport in Great Britain and Germany (Sjöblom, 2007, based on a much more extensive dissertation from the same year) and a brief comparison of the early road motorization in France, Germany, and Switzerland in an article about early car racing (Merki, 2002, based on a much more extensive book publication from the same year, in German). Immo Sievers analyzed the early connections between the British and German car industries (1995, in German), following in the tracks of the 'European John Rae,' James M. Laux and his study of the European car industry (1992). A comparison between the early bicycle cultures in the Netherlands and Germany comes from Anne-Katrin Ebert (2004, prelude to a much more extensive dissertation from 2009, in German).

Works on National Automobilism

By far most publications on *road motorization* are *national* in scope.[2] American classics are John Rae (1971, 1984, emphasizing production rather than consumption) and James Flink (on the emergence of car culture, 1970, and his last monograph, 1988, both innovative because of their use of trade and club journals as historical source). In Flink's footsteps is Rudi Volti's cultural history of the 'technology' of the car from 2004. All industrialized countries got their pioneering historians, sometimes modeled after Flink's approach, sometimes going beyond his method. Australia's pioneering study of national car culture is more modeled after Flink (Davison, 2004), but a quite original classic history of road transport *policy* in the UK is William Plowden (1971). A more recent excellent entrance into the cultural history of American automobilism is Seiler (2008) but see the previous chapter for several more fine studies. Of the several national histories of France the one by Jean Orselli is the most recent and the most comprehensive, emphasizing road unsafety (2011, based on a much more extensive dissertation, in three volumes, 2009, all in French). Mathieu Flonneau is one of the French pioneers of autohistory-related New Mobility Studies with several publications, from a dissertation (see further down, under urban histories) via a polemical treatise about scholarly anti-automobilism (2010, in French) to more popular histories on traffic signs. Flonneau's no less productive counterpart for Germany is the cultural historian Kurt Möser, who not only wrote a history of German car culture (2002), but also connected this with the culture of flying (2009, both in German) and even boating. More focused on production is Rainer Flik (2001, in German). An early political-economic historical analysis comes from Krämer-Bodoni, Grymer, and Rodenstein (1971, in German). See Mom and Filarski, mentioned above, for an example on Dutch automobilism.

The pioneer on Soviet automobilism (2008) and on the Eastern bloc in general (2011) is Lewis Siegelbaum, but he also published on racing and rallying during the Interbellum (2005), as well as migration (Siegelbaum and Moch, 2014). The history of Indian road mobility (especially freight and passenger transport) can be found in Stefan Tetzlaff's contribution in the edited volume by Arve Hansen and Kenneth Bo Nielsen (2017), based on his dissertation from 2015. In Latin America three countries have their historian of automobile culture: Joel Wolfe for Brazil (2010), Melina Piglia for Argentina (2014, in Spanish), and Rodrigo Booth's dissertation on early automobilism in Chile (2009, in Spanish). Rodrigo Booth and Melina Piglia (2009) mention other recent publications for Latin America (and, again, in general the *Mobility in History* series, first published by Alphil in Switzerland, later by Berghahn Books, are useful sourcebooks). A study of Mexican road transport between 1926 and 1960 is the dissertation by Benjamin Fulwider (2009). The pioneering study for Africa is the edited volume by Jan-Bart Gewald, Sabine Luning, and Klaas van Walraven (2009). There are many historical studies on colonial transport in Africa, often written by anthropologists, for instance an early analysis in Ghana by K. B. Dickson (1961), E. K. Hawkins's study on Uganda's road transport history (1962), or Rolf Hofmeier's study of Tanzania (1973). A recent exemplary study of the latter country comes from Joshua Grace (2021). For a later study about the impact of road and rail transport in colonial Mozambique see Arlindo Gonçalo Chilundo's dissertation (1995). A wrap-up of a century of African road transport comes from Gina Porter (2012).

Periods of Automobilism

Many publications on the history of automobilism focus on certain periods or aspects. As to the *periods*, the *emergence* of national car cultures is well represented. Joachim Radkau places Germany's fledgling automobilism in the culture of nervosity of the fin de siècle (1991, 1994, both in German), whereas Catherine Bertho Lavenir brings the many parallels with the early bicycle culture and tourism to the fore (1999, in French). Uwe Fraunholz wrote by far the most extensive thesis on early anti-car resistance in Germany (2002, in German), while Gijs Mom traces efforts by road engineers to improve safety (2007). Clay McShane and Joel Tarr place the American car in the tradition of horse transport (2007). Scholarly analyses of colonial fledgling car cultures are rare, with Stéphanie Ponsavady's published dissertation (2018) being one of the best.

Several analyses of and narratives on the so-called 'coordination crisis' (the struggle between rail and road) have their origin in the *Interbellum*, such as the popular history of American coordination by Stephen B. Goddard (1994). A good overview is provided by Gérard Duc et al. (2014). The post-World War II *exuberance period* received some analyses by Gijs Mom (this volume), and by Mathieu Flonneau for France (2016, in French). Most of them are national in scope.

Gender, Safety, Justice

Pioneers in the study of *gender aspects* of automobilism are Virginia Scharff (1991), Margaret Walsh (2015), and Georgine Clarsen (2008) (see also Chapter 1). Lately, gender aspects are included in studies of inequality within *mobility justice*, pioneered, among others, by Mimi Sheller (2018). For a feminist approach see Jennie C. Stephens (2024) (see also Chapter 5).

The history of *road safety* became extra important recently in relation to the question of the future of automobilism (see Chapters 4 and 5), but was pioneered more than a quarter century ago by Daniel M. Albert in his unpublished dissertation (1997) on racist aspects of road safety, ten years later accompanied by Jameson M. Wetmore (2008), focusing on car technology, Lee Vinsel's dissertation (2011) on safety policy and regulation, and Arwen Mohun (2013, emphasizing industrial safety and risk), all three coming from the field of the history of technology. A more epidemiological approach has been pioneered for the UK by Bill Luckin (1997) as a 'war on the road.' For France, more recently, Alice Milor focuses on European policy (2019). A transnational approach was pioneered by Stève Bernardin and Harrison Grafos (2008).

Remarkably, mobility history is poor in *environmental histories*, rare recent ones being Tom McCarthy (2007) and Christopher W. Wells (2012). Conversely, environmental histories are poor on mobility. For American general environmental policy see for instance Richard N. L. Andrews (1999). Frederick Buell wrote a history of American environmentalism informed by belletristic literature (2004, 2005). On its relation to American suburban nature: Christopher C. Sellers (2012). On the American 'wilderness movement' and automobilism: Paul S. Sutter (2002) and on 'windshield wilderness': David Louter (2006). Rudi Volti wrote an early article on California's pollution abatement policy (2003); of the many studies on this state's ZEV (Zero Emission Vehicle) mandate see for instance Joeri Wesseling's dissertation (2015) on the industry's responses to the mandate. On British environmentalists' protests against road-building see Alfred W. Crosby (1988). A pioneer in the history of German environmentalism is Frank Uekötter (2014), including his contribution on car pollution in E. Melanie DuPuis's edited volume of 2004. On Germany's nature and landscape protection policy see the edited volume by Thomas Lekan and Thomas Zeller (2005). On Italy's environmental aspects of automobilism: Federico Paolini (2012). A must-read on environmental and safety aspects of urban transport in Latin America is urban planner Eduardo Alcântara Vasconcellos (2001). On Latin American environmental policy in general see Shawn William Miller (2007) and Mark Carey (2009). On Delhi's environmental policy: Amita Baviskar (2011). On China's environmental policy see Maohong Bao (2010), and during Mao's reign: Judith Shapiro (2001). On Western 'ecological imperialism' see Alfred W. Crosby (1988).

Touring, Tourism, Travel

A pioneer in international *tourism, travel,* and rail transport is Laurent Tissot (2003; on the European industry: 2000, in French), followed by a transnational approach in the edited volume by Eric G. E. Zuelow (2011); likewise, European automotive tourism in Gijs Mom (2011). On nineteenth- and early twentieth-century tourism and belletristic literature see James Buzard (1993). On the origins of sightseeing: Judith Adler (1989). For a comparison between German, French, and British tourism between 1930 and 1980 see Christopher Kopper (2009, in German). For a broader 'history of leisure' see Rudy Koshar (2002). A pioneer in the history of tourism and 'vacationing' is Orvar Löfgren (1999).

Classics about early *car touring* in the *US* are Warren James Belasco (1979), Peter J. Hugill (1985) and John A. Jakle (1981). A pioneer in American pre-World War II tourism history is Marguerite S. Shaffer (2001). Tourism in the American West (including the national parks) is examined in the edited volume by David M. Wrobel and Patrick T. Long (2001) and early car tourism in California by Andrew Gross (2003).

On *British, Australian, and Canadian* nostalgic 'battlefield tourism' during the Interbellum see David W. Lloyd (1998). A pioneer on *British tourism* is John K. Walton (2000, 2005), and on working-class tourism see Susan Barton (2005). An early edited volume on travel in *Germany* is Dieter Kramer and Ronald Lutz (1992). German war tourism: Charlotte Heymel (2007, in German). Early German car tourism to Switzerland:

Uwe Fraunholz (2002, in German). Tourism in the Third Reich: Peter J. Brenner (1997, in German), Shelley Baranowski (2004), and Kristin Semmens (2005); emergence of German mass tourism: Hasso Spode (2009, in German).

Early bicycle tourism in *France* is examined for instance in the edited volume by Jean-Denys Devauges and Didier Masseau (1997, in French) and in the *UK* by Gary Allen Tobin (1974). French early tourism by car, during the Interbellum (related to the nation and empire) and from the 1930s to the 1970s: Ellen Furlough (1998; 2002, the latter in French). For a study on the postwar revival of tourism in France through the Marshall Plan: Brian A. McKenzie (2003), and on 'Cold War holidays' by Americans: Christopher Endy (2004). Tourism to *colonial Algeria*: Colette Zytnicki (2013, in French). Mass tourism in the *Mediterranean*: Luciano Segreto, Carles Manera, and Manfred Pohl (2009). Early travel writing in the *Soviet Union*: Andreas Renner (2011); tourism during late Stalinism: Anne E. Gorsuch (2003). Car tourism in *Romania*: Adelina Oana Stefan (2015). Car tourism in pre-independence *Africa*: Gordon Pirie (2013). Tourism before World War II in *China*: Yajun Mo (2021). Travel in the *'Third World'*: David L. Gladstone (2005).

Systemic Aspects

Regarding other *aspects* of histories, the oldest histories of automobilism focused on *production* (see, for instance, Rae and Laux above), including early studies of the transnational network-forming of the industry, for instance by Mira Wilkins and Frank Ernest Hill on Ford's spread over the globe (1964). Soon car *consumption* became included (see Flink above), lately added to with a rethinking of the early car buyers (middle-class or working-class?) by Craig Horner (2021) (see Chapter 1). On the history of car *design* see for instance David Gartman (1994). Another part of car consumption history is the tinkering and professional and amateur *maintenance* aspect, pioneered in the US by historians of technology such as Stephen L. McIntyre and his article on the American repair industry (2000, based on his dissertation from 1995), and Kathleen Franz (2005) and Kevin L. Borg (2007) on tinkering and professional mechanics. Stefan Krebs started a German research tradition on this aspect (2012, in German). See Chapter 4 on the special role of car modifying for groups of young ethnic tinkerers, and for the role of repair as creative practice in Southern Europe see Alexia Sodia Papazafeiropoulou (2012), in Eastern Europe see Möser (2011), and for the Global South see for instance Sara S. Berry (1983) or, more recently, Joshua Grace (2021) (see Chapter 3). On the transnational history of automotive *technology* see Mom (2014); on the history of an important car material, enabling historian and sociologist Mimi Sheller to connect car history to colonization, extractive practices, and 'mobility justice', see her *Aluminum Dreams* (2014).

Roadbuilding

Another *systemic aspect* is the vast subfield of roadbuilding history, pioneered by the dissertation of Massimo Moraglio for Italian freeways (2007, in Italian), Thomas Zeller for the German *autobahnen* (2002, in German), Peter Merriman for the British M1 motorway (2007), and Gijs Mom for the Netherlands (2007) and for a more general history of roadbuilding technology (2004).

But there are predecessors, such as Arnulf Grübler's study of the 'rise and fall of infrastructures' in transport (1990). A pioneer for Eastern Europe is the very recent study by Lyubomir Pozharliev (2023); for Russia's 'roadlessness' see Lewis H. Siegelbaum (2008). American pioneers are historians of technology Mark H. Rose (1979) on the Interstate system and Bruce E. Seely on American road and highway building in general (1987), and Michael R. Fein on the state of New York (2008). An old American classic is Lewis Mumford on highways and cities (1964). Michael K. Bess provided an overview of Latin American roadbuilding in 2017 and contributed his dissertation on Mexico from the same year. A recent revival of road research was initiated by anthropologists such as Dimitris Dalakoglou and Penny Harvey in the early 2010s; they edited a book on this topic in 2015. In this tradition see on Asia: Luke Heslop and Galen Murton (2021).

Urban Automobilism

The American pioneer of *urban transport histories* is no doubt Clay McShane (1994). For France it is Mathieu Flonneau with his Parisian history of automobilism (2005, in French, based on a three-volume dissertation).

Many earlier histories are dedicated to one or two cities, and are tributary to the history of technology (and are then combined with histories of other urban technologies, such as sewers, or roadbuilding). A good overview for Europe is given by Michael Hård and Thomas J. Misa (2003); an early American example of urban transport history is Paul Barrett's treatise on Chicago's struggle between auto and transit (1983).

An analysis of the Dutch shift from a national to an 'urban view' in the history of transport planning is attempted by Mom (2016).

A classic of urban rickshaw history is James Warren's analysis of Singapore's transport culture (2003).[3] For the study of Asian urban transport history the studies by Howard Dick and Peter J. Rimmer are a necessary starting point, for instance Rimmer's *Rikisha to Rapid Transit* from 1986, or Dick and Rimmer's overview update from 2003.

More recent studies of this very wide field use individual cities as *pars pro toto* of the national culture, and they are certainly not limited to the West. Shawn Miller's recent history of Rio de Janeiro's streets (2018) is an example written in the tradition of McShane, but Govind Gopakumar's history of urban congestion in Indian cities (2020) and Ritajyoti Bandyopadhyay's work on twentieth-century riots and the post-World War II struggle between the car and street hawkers in Calcutta (2022) are examples of the explanatory power when history writing is inspired by social science from within the 'mobility turn.'[4]

A special subfield in urban mobility history regards the *taxi*, for instance in New York by Graham Russell Gao Hodges (2007), in prewar Paris by Mathieu Flonneau (2011, in French), or in early Amsterdam (including the history of horse cabs and electric propulsion) by Mom (2003). On illegal taxis (jitneys) often driven while black in the US during the 1920s: Blaine A. Brownell (1972) and in Canada: Donald F. Davis (1989). See Chapters 2, 3, and 4 for the many motorcycle and bicycle taxis, as well as the rickshaws in the Global South.

Other Mobilities

Histories on *walking* and pedestrianism are rare. Barbara Schmucki (2012) is one of the first. Histories of *cycling and its culture* in its own right (see above for cycling as a cultural and technical predecessor of automobilism) abound, but one of the best early ones is a social history by Rüdiger Rabenstein (1991, in German). Glen Norcliffe wrote a comparable study on Canadian early cycling 'toward modernity' (2001). Gender aspects of early cycling are dealt with by Philip Gordon Mackintosh and Glen Norcliffe (2011). Frédéric Héran wrote a history of nearly two centuries for France (2014). Ruth Oldenziel and Adri Albert de la Bruhèze published a study of European urban bicycle path construction (2011), Georgine Clarsen documented Australian cycling culture (2015), while Xu Tao wrote China's early cycling history in a Chinese dissertation, and a summary of its more commercial sides appeared in English (2013).[5]

On cycling's contribution to Namibia's economy in the 1930s see Anthony I. Nwabughuogu (1984). Of special interest are cycling travelogues, especially international ones, such as Dervla Murphy's trip from Ireland to India (2010). A volume stressing the sustainable characteristics of cycling has been edited recently by Martin Emanuel, Frank Schipper, and Ruth Oldenziel (2020).

Motorcycle histories are scarce (individual histories of its culture are to be found in the *International Journal of Motorcycle Studies*), one of the pioneers being Klaus Benesch (2013). On the role of motorcycle racing in Nazi Germany see Adelheid von Saldern (1992). On early motorcycle culture in the US see the dissertation by Sammy Kent Brooks (1975), and for the gendering of this culture in Germany see Sasha Disko (2016). The cultural and social history of *scooters* and *mopeds* still has to be written.

Electric propulsion is as old as internal combustion cars, according to Gijs Mom's and David Kirsch's dissertations, the latter focusing on the US (2000), the former providing an international comparison (2004, a translation from the dissertation, in Dutch, from 1997) (see Chapter 1). Both their works were preceded by popular technical histories such as the two, one on pure electric and the other on hybrid vehicles, by Ernest H. Wakefield (1994 and 1998) and by an anthropological study by Michael Brian Schiffer (1994). An early study focusing on Germany is Dietmar Abt (1998, in German). On California's attempts to introduce electric traction in the 1990s see Mark B. Brown (2001). For an analysis of similar French attempts from the viewpoint of actor network theory see Michel Callon (1986).

On the history of *steam and diesel cars* see M. Hård and A. Jamison (1997). A comparison between the diesel car's history in the US and Germany in Christopher Neumaier (2010, based on his dissertation from the same year, in German). On *alcohols* as alternative fuel see, for methanol in Sweden, Erland Mårald (2010), and, more in general, Hal Bernton, William Kovarik, and Scott Sklar (2010, an improved edition over the previous one from 1982). More in general about *alternatives to the fossil-fueled car* see Mikael Hård's edited volume from 1992. See also Andreas Knie's sociological histories of the *diesel engine* (1991) and the *Wankel engine* (1994, both in German).

The history of *freight transport on roads* and the *trucking industry* is extensive but scattered. On freight transport from a transatlantic perspective see Mom (2014). On the history of the industry, also in Europe, and the so-called 'coordination crisis' (the struggle between rail and road) see the edited volume by Christopher Kopper and Massimo Moraglio (2015). From the perspective of the European railroads: Laurent Tissot (1998). For the coordination history of France: Nicolas Neiertz (1999). An early study of the military standardization of trucks in the American army is Marc K. Blackburn (1996, based on his dissertation from 1992). On the controversial history of what some have called the 'Snell hoax' (about the 'conspiracy' by General Motors to 'kill' rail transport), see Nicholas Bloom (2023). On the emergence of early federal regulation see William R. Childs (1985). On postwar trucking in the US: Shane Hamilton (2008). A constructivist approach of the American truck's technical history is Louis Rodriquez's dissertation from 1997. Dawn Chatty wrote a study of the development from camel to truck among the Bedouins (2013; the first edition is from 1986).

Classics on British interwar bus regulation are Corinne Mulley (1983) and Stephen Glaister and Corinne Mulley (1983); an American pioneer on the same theme is John B. Rae (1973; first edition 1967), later accompanied by Margaret Walsh (2000). The motor bus was also subject to the 'coordination crisis,' for instance in the US, from a business-history perspective: Gregory Lee Thompson (1993).

Representations

Literary representations of automobilism abound, the pioneers being Paul Fussell on British 'literary travel' (in this case between the war) (1980) and Karen R. Lawrence (1994) on women's part therein. Early American pioneers are Janis P. Stout (1983) and Cecilia Tichi (1987), and of a revived genre dedicated exclusively to car culture Ronald Primeau (1996, and a publication on women on the road, 1999), Roger N. Casey (1997), Stout (1998, on women's narratives), Katie Mills (2006, based on her dissertation from 2000), and Deborah Clarke (2007, on women). More in general, as a part of popular culture: Nicholas Zurbrugg (1998). For travel representation in feminist fiction see Heidi Slettedahl Macpherson (2000). For Germany: Dorit Müller (2011, based on her dissertation from 2004, in German) and Anita Perkins (2016). On more general travel writing and the novel, for instance: Percy G. Adams (1983), or on travel by Japanese in China: Joshua A. Fogel (1996). Representation of German car culture (on the autobahn) in literature and film can be found in the edited volume by Jan Röhnert (2014), but literature and film are also combined in Katie Mills's publication (including television) and Dorit Müller's dissertation, mentioned above, as well as Siegfried Reinecke's earlier study (1992, in German) or the somewhat forgotten study by Lyn Elizabeth Elliot (2000). For a theory about the relation between literature, film, and automobilism see Charles Grivel's contribution to the edited volume by Jochen Mecke and Volker Roloff (1999). For representations of walking in literature and film in early German cinema see Anke Gleber (1999).

As prolific is the genre of *film studies* and automobilism, American and British pioneers being David Laderman (2002) and Iain Borden (2013). They are all successors of Kenneth Hey's early study from 1980. 'Auto-erotics' in very early American films are to be found in Jennifer Parchesky (2006). On American cinema and travel in general see the edited volume by Jeffrey Ruoff (2006). Yvonne Tasker's edited volume focuses on action cinema (2004). For French cinema and the car see Charles Grivel (1999). On the *road movie* for instance (among many studies) see Ron Eyerman and Orvar Löfgren (1995). See also the edited volume by Steven Cohan and Ina Rae Hark (1997).

Notes

1. See the Bibliography for the complete references.
2. Several of the national histories mentioned here are reviewed in more detail in the Introduction to the first part of my trilogy (2015).
3. For more sources on other Asian cities see this volume, Chapters 2 and 4.
4. For a critical review of these and some other recent studies see Gijs Mom, "'Our City is Ruined: Reason: Road Widening': A Review Essay on New Mobility Studies, Social Science, and History," *Transfers* 12, no. 3 (Winter 2022): 95–105.
5. For more sources see Chapter 1.

Bibliography

Note: Titles with an asterisk (*) are mentioned in the Annotated Canon.

*Abt, Dietmar. *Die Erklärung der Technikgenese des Elektroautomobils* (Frankfurt am Main: Lang, 1998) (European University Studies, series V, vol. 2295).

*Adams, Percy G. *Travel Literature and the Evolution of the Novel* (Lexington, Kentucky: The University Press of Kentucky, 1983).

*Adas, Michael. *Machines as the Measure of Men: Science, Technology, and Ideologies of Western Dominance* (Ithaca, NY: Cornell University Press, 1989).

*Adler, Judith. "Origins of Sightseeing," *Annals of Tourism Research* 16 (1989): 7–29.

Agbiboa, Daniel E. "Introduction: Transport, Transgression and Politics in African Cities," in Daniel E. Agbiboa (ed.), *Transport, Transgression and Politics in African Cities: The Rhythm of Chaos* (Abingdon: Routledge, 2019) (Cities and Society, ed. Chris Pickvance), 1–16.

——. "State Law as a Means of Resistance: Okada Riders versus the Lagos State Government," in Daniel E. Agbiboa (ed.), *Transport, Transgression and Politics in African Cities: The Rhythm of Chaos* (Abingdon: Routledge, 2019) (Cities and Society, ed. Chris Pickvance), 147–71.

——. "Stomach Infrastructure: Informal Transport, Electoral Politics and the Precariousness of Patronage in Lagos," in Daniel E. Agbiboa (ed.), *Transport, Transgression and Politics in African Cities: The Rhythm of Chaos* (Abingdon: Routledge, 2019) (Cities and Society, ed. Chris Pickvance), 43–59.

Aguiar, Marian. *Tracking Modernity: India's Railways and the Culture of Mobility* (Minneapolis: University of Minnesota Press, 2011).

Akrich, Madeleine. "The De-Scription of Technical Objects," in W. E. Bijker and J. Law (eds.), *Shaping Technology, Building Society* (Cambridge, MA: MIT Press, 1992), 205–23.

Alacevich, Michele. *Visualizing Uncertainties, or How Albert Hirschman and the World Bank Disagreed on Project Appraisal and Development Approaches* (n.p. [New York]: World Bank, Information Management and Technology Network, Knowledge and Information Service Unit, November 2012).

*Albert, Daniel M. "Order out of Chaos: Automobile Safety, Technology and Society 1925 to 1965," dissertation (Ann Arbor: University of Michigan, 1997).

*Aldcroft, D. H., and M. J. Freeman. *Transport in the Industrial Revolution* (Manchester: Manchester University Press, 1983).

Allen, J. W. *Wheel Magic or Revolutions of an Impressionist* (London: John Lane, n.d. [1909]).

Alvarez León, Luis F., and Yuko Aoyama. "Industry Emergence and Market Capture: The Rise of Autonomous Vehicles," *Technological Forecasting & Social Change* 180 (2022): 121661.

Amburn, Ellis. *Subterranean Kerouac: The Hidden Life of Jack Kerouac* (New York: St. Martin's Press, 1998).

Anderson, Ben, and Paul Harrison. "The Promise of Non-Representational Theories," in Ben Anderson and Paul Harrison (eds.), *Taking-Place: Non-Representational Theories and Geography* (Farnham: Ashgate, 2010), 1–34.

Anderson, G. L. "The Social Economy of Late-Victorian Clerks," in Geoffrey Crossick (ed.), *The Lower Middle Class in Britain 1870–1914* (London: Croom Helm, 1977), 113–33.

*Andrews, Richard N. L. *Managing the Environment, Managing Ourselves: A History of American Environmental Policy* (New Haven, CT: Yale University Press, 1999).

Appleyard, Bryan. *The Car: The Rise and Fall of the Machine That Made the Modern World* (London: Weidenfeld & Nicolson, 2022).

Apter, Emily. *Against World Literature: On the Politics of Untranslatability* (London: Verso, 2013).

Arnold, David. "Subaltern Streets: India, 1870–1947," in Tariq Jazeel and Stephen Legg (eds.), *Subaltern Geographies* (Athens: University of Georgia Press, 2019), 36–57.

Arnold, David, and Erich DeWald. "Cycles of Empowerment? The Bicycle and Everyday Technology in Colonial India and Vietnam," *Comparative Studies in Society and History* 53, no. 4 (2011): 971–96.

Aronson, Sidney H. "The Sociology of the Bicycle," *Social Forces* 30 (1951–52): 305–12, reprinted under the same title in: Marcello Truzzi (ed.), *Sociology and Everyday Life* (Englewood Cliffs, NJ: Prentice-Hall, 1968), 293–303.

Arrighi, Giovanni. *The Long Twentieth Century: Money, Power, and the Origins of Our Times* (London: Verso, 2010 [1994]).

Athukorala, Prema-Chandra, and C. Veeramani. "From Import Substitution to Integration into Global Production Networks: The Case of the Indian Automobile Industry," *Asian Development Review* 36, no. 2 (2019): 72–99.

Attardo, Salvatore. "Irony as Relevant Inappropriateness," *Journal of Pragmatics* 32 (2000): 793–826.

Badenoch, Alexander. "Touring between War and Peace: Imaging the 'Transcontinental Motorway,' 1930–1950," *Journal of Transport History* 28, no. 2 (September 2007): 192–210.

Baca, George. "Neoliberal Narratives of Crisis: The Feeble Cries of a Vanishing 'Class,'" *Dialectical Anthropology* 41 (2017): 377–85.

*Bagwell, Philip, and Peter Lyth. *Transport in Britain: From Canal Lock to Gridlock* (London: Hambledon and London, 2002).

Bailey, Peter. "White Collars, Gray Lives? The Lower Middle Class Revisited," *Journal of British Studies* 38 (July 1999): 273–90.

Bailyn, Bernard. *Atlantic History: Concept and Contours* (Cambridge, MA: Harvard University Press, 2005).

Bakhtin, Mikhail. *Rabelais and His World*, trans. Helene Iswolsky (Bloomington: Indiana University Press, 1984 [1965]).

*Baldasseroni, Louis, Étienne Faugier, and Claire Pelgrims (eds.). *Histoire des transports et des mobilités en France: XIX*[e]*–XXI*[e] *siècles* (Malakoff: Armand Colin, 2022) (Collection U: Histoire).

Balkmar, Dag. "On Men and Cars: An Ethnographic Study of Gendered, Risky and Dangerous Relations," unpublished doctoral dissertation (Linköping: Linköping University, 2012).

Ballard, J. G. *Miracles of Life: Shanghai to Shepperton; An Autobiography* (London: Fourth Estate, 2008).

Balzer, Harley. "Russia's Middle Classes," *Post-Soviet Affairs* 14, no. 2 (1998): 165–86.

*Bandyopadhyay, Ritajyoti. *Streets in Motion: The Making of Infrastructure, Property, and Political Culture in Twentieth-Century Calcutta* (Cambridge: Cambridge University Press, 2022).

*Bao, Maohong. "The Evolution of Environmental Problems and Environmental Policy in China," in J. R. McNeill and Corinna R. Unger (eds.), *Environmental Histories of the Cold War* (Washington, DC/Cambridge: German Historical Institute/Cambridge University Press, 2010), 323–40.

*Baranowski, Shelley. *Strength through Joy: Consumerism and Mass Tourism in the Third Reich* (Cambridge: Cambridge University Press, 2004).

*Barker, Theo (ed.). *The Economic and Social Effects of the Spread of Motor Vehicles: An International Centenary Tribute* (London: MacMillan, 1987).

*Barker, Theo, and Dorian Gerhold. *The Rise and Rise of Road Transport, 1700–1990* (Houndmill: Manchester University Press, 1993) (Studies in Economic and Social History).

*Barrett, Paul. *The Automobile and Urban Transit: The Formation of Public Policy in Chicago, 1900–1930* (Philadelphia: Temple University Press, 1983) (Technology and Urban Growth, eds. Blaine Brownell, Mark S. Foster, Zane L. Miller, Mark Rose, and Howard J. Sumka).

*Barton, Susan. *Working-Class Organisations and Popular Tourism, 1840–1970* (Manchester: Manchester University Press, 2005) (Studies in Popular Culture, ed. Jeffrey Richards).

Bassnett, Susan. "Travel Writing and Gender," in Peter Hulme and Tim Youngs (eds.), *The Cambridge Companion to Travel Writing* (Cambridge: Cambridge University Press, 2002), 225–41.

Bauman, Zygmunt. *Liquid Life* (Cambridge: Polity Press, 2006 [2005]).

*Baviskar, Amita. "Cows, Cars and Cycle-Rickshaws: Bourgeois Environmentalists and the Battle for Delhi's Streets," in Amita Baviskar and Raka Ray (eds.), *Elite and Everyman: The Cultural Politics of the Indian Middle Classes* (London: Routledge, 2011), 391–418.

Bay, Mia. *Traveling Black: A Story of Race and Resistance* (Cambridge, MA: Belknap Press of Harvard University Press, 2021).

Bedi, Tarini. *Mumbai Taximen: Autobiographies and Automobilities in India* (Seattle: University of Washington Press, 2022).

*Belasco, Warren James. *Americans on the Road: From Autocamp to Motel, 1910–1945* (Cambridge, MA: MIT Press, 1979).

*Benesch, Klaus. "*Our Bikes Are Us*: Speed, Motorcycles and the American Tradition of a 'Democratic' Technology," in: id. (ed.), *Culture and Mobility* (Heidelberg: Universitätsverlag Winter, 2013) 65–80.

Berking, Helmuth. "Raumtheoretische Paradoxien im Globalisierungsdiskurs," in Helmuth Berking (ed.), *Die Macht des Lokalen in einer Welt ohne Grenzen* (Frankfurt: Campus, 2006), 7–22.

*Bernardin, Stève, and Harrison Grafos. "Making the Automobile Safe for the World? A Transnational Perspective on Car Safety Standards, 1958–1998," *Entreprises et histoire* 51 (2008): 69–87.

*Bernton, Hal, William Kovarik, and Scott Sklar. *The Forbidden Fuel: A History of Power Alcohol; New Edition* (Lincoln: University of Nebraska Press, 2010).

*Berry, Sara S. "From Peasant to Artisan: Motor Mechanics in a Nigerian Town" (Boston: African Studies Center, 1983) (Working Paper No. 76).

*Bertho Lavenir, Catherine. *La roue et le stylo: Comment nous sommes devenus touristes* (Paris: Editions Odile Jacob, April 1999).

*Bess, Michael K. "On the Course of 'Progress': A Review of Literature on Road Building in Latin-America," *Mobility in History* 8 (2017): 35–41.

*——. *Routes of Compromise: Building Roads and Shaping the Nation in Mexico, 1917–1952* (Lincoln: University of Nebraska Press, 2017) (The Mexican Experience, ed. William H. Beezley).

Bi, Marie Richard Zouhoula. "The Woro-Woro *Gares Routières* of Abidjan: Artisanal Transport and Local Governance in Côte d'Ivoire's Largest City," *Africa Today* 65, no. 2 (Winter 2018): 22–33.

Bianciardi, Luciano. *La Vita Agra or It's a Hard Life* (London: Hodder and Stoughton, 1965 [1962]).

Bieker, Georg. "A Global Comparison of the Life-Cycle Greenhouse Gas Emissions of Combustion Engine and Electric Passenger Cars" (Berlin: ICCT, 2021).

Bijker, Wiebe Eco. *The Social Construction of Technology* (Eijsden [PhD thesis University Twente, 1990).

Bijsterveld, Karin, Eefje Cleophas, Stefan Krebs, and Gijs Mom. *Sound and Safe: A History of Listening behind the Wheel* (Oxford: Oxford University Press, 2014).

Bishop, Ted. "*Tempo Giusto*: The Art of the Slow Ride," in Klaus Benesch (ed.), *Culture and Mobility* (Heidelberg: Universitätsverlag Winter, 2013), 81–88.

Black, Edwin. *Internal Combustion: How Corporations and Governments Addicted the World to Oil and Derailed the Alternatives* (New York: St. Martin's Press, 2006).

*Blackburn, Marc K. *The United States Army and the Motor Truck: A Case Study in Standardization* (Westport, CT: Greenwood Press, 1996) (Contributions in Military Studies, 163).

*Bloom, Nicholas Dagen. *The Great American Transit Disaster: A Century of Austerity, Auto-Centric Planning, and White Flight* (Chicago: University of Chicago Press, 2023) (Historical Studies of Urban America, eds. Lilia Fernández et al.).

Boatcă, Manuela. "Semiperipheries in the World-System: Reflecting Eastern European and Latin American Experiences," *Journal of World-Systems Research* 12, no. 11 (December 2006): 321–46.

Böhm, Steffen, Campbell Jones, Chris Land, and Mat Paterson. "Introduction: Impossibilities of Automobility," in Steffen Böhm et al. (eds.), *Against Automobility* (Malden, MA: Blackwell, 2006), 3–16.

Bolaño, Roberto. *The Savage Detectives*, trans. Natasha Wimmer (London: Picador, 2009 [1998]).

*Booth, Rodrigo. "Automóviles y Carreteras: Movilidad, modernización y transformación territorial en Chile, 1913–1931," unpublished dissertation (Santiago: Pontificia Universidad Católica de Chile, September 2009).

*Booth, Rodrigo, and Melina Piglia. "New Developments in a Neglected Field: Transport and Mobility in Latin American Recent Historiography," in Gijs Mom, Gordon Pirie, and Laurent Tissot (eds.), *Mobility in History: The State of the Art in the History of Transport, Traffic and Mobility* (Neuchâtel: Alphil, 2009), 159–65.

*Borden, Iain. *Drive; Journeys through Film, Cities and Landscapes* (London: Reaktion Books, 2013).

*Borg, Kevin L. *Auto Mechanics: Technology and Expertise in Twentieth-Century America* (Baltimore: Johns Hopkins University Press, 2007) (Studies in Industry and Society, ed. Philip B. Scranton).

Borowy, Iris. "'To Study "Accidents" as a Whole May Be Like Studying "Disease" as a Whole': The Epidemic of Road Traffic Injuries," in Jörg Vögele, Stefanie Knöll, and

Thorsten Noack (eds.), *Epidemics and Pandemics in Historical Perspective* (Herbolzheim: Centaurus Press, 2015), 177–189.

Bosco, Fernando J. "Actor-Network Theory, Networks, and Relational Approaches in Human Geography," in Stuart Aitken and Gill Valentine (eds.), *Approaches to Human Geography* (London: Sage, 2006), 136–46.

Bose, Rajat, and P. K. Sarkar. "The Role of Cycle Rickshaws—Low Cost Means of Transport in Old Part of a City," in X. Godard and I. Fatonzoun (eds.), *Urban Mobility for All/La Mobilité Urbaine pour Tous: Proceedings of the Tenth International CODATU Conference, Lomé, Togo, 12–15 November 2002* (Lisse: A. A. Balkema, 2002), 199–203.

Boutle, Ian. "'Speed Lies in the Lap of the English': Motor Records, Masculinity, and the Nation, 1907–14," *Twentieth Century British History* 23, no. 4 (2012): 449–72.

Bramall, Chris. *Chinese Economic Development* (London: Routledge, 2009).

Brand, Christian, Jillian Anable, Ioanna Ketsopoulou, and Jim Watson. "Road to Zero or Road to Nowhere? Disrupting Transport and Energy in a Zero Carbon World," *Energy Policy* 139 (2020): 111334.

Braun, Robert, and Richard Randell. *Post-Automobility Futures: Technology, Power, and Imaginaries* (Lanham, MD: Rowman and Littlefield, 2022).

——. "Towards Post-Automobility: Destituting Automobility," *Applied Mobilities* 8, no. 3 (2023): 201–17.

*Brenner, Peter J. "Schwierige Reisen: Wandlungen des Reiseberichts in Deutschland 1918–1945," in Peter J. Brenner (ed.), *Reisekultur in Deutschland: Von der Weimarer Republik zum "Dritten Reich"* (Tübingen: Max Niemeyer, 1997), 127–76.

Brogan, Una. *The Alternative Modernity of the Bicycle in British and French Literature, 1880–1920* (Edinburgh: Edinburgh University Press, 2024 [2022]) (Edinburgh Critical Series in Victorian Culture).

——. "Cycling and Narrative Structure: H.G. Wells's *The Wheels of Chance* and Maurice Leblanc's *Voici des ailes*," in Marian Aguiar, Charlotte Mathieson, and Lynne Pearce (eds.), *Mobilities, Literature, Culture* (Cham: Palgrave Macmillan, 2019), 237–57.

——. "Domesticated Mobility: Exploring the Limits of Women's Liberation through Cycling in British Literature, Guide Books and Periodicals, 1890–1910," *Cycle History 29: Proceedings of the 29th International Cycling History Conference* (London, 2018): 55–60.

——. "Liberation on Two Wheels: Social Change and the Bicycle in H.G. Wells's *Kipps* and *The History of Mr. Polly*," *The Wellsian* 41 (2018): 5–27.

*Brooks, Sammy Kent. "The Motorcycle in American Culture: From Conception to 1935," unpublished dissertation (Washington, DC: George Washington University, 1975).

*Brown, Mark B. "The Civic Shaping of Technology: California's Electric Vehicle Program," *Science, Technology, & Human Values* 26, no. 1 (Winter 2001): 56–81.

*Brownell, Blaine A. "The Notorious Jitney and the Urban Transportation Crisis in Birmingham in the 1920's [sic]," *Alabama Review* 25 (April 1972): 105–18.

*Buell, Frederick. *From Apocalypse to Way of Life: Environmental Crisis in the American Century* (New York: Routledge, 2004 [2003]).

*Buell, Lawrence. *The Future of Environmental Criticism: Environmental Crisis and Literary Imagination* (Malden, MA: Blackwell, 2005).

Burawoy, Michael. "Review: The End of Sovietology and the Renaissance of Modernization Theory," *Contemporary Sociology* 21, no. 6 (November 1992): 774–85.

Burgers, Christian, and Margot van Mulken. "Het ironisch spectrum: Een overzicht van onderzoek naar het begrip en de retorische effecten van verbale ironie," *Tijdschrift voor Taalbeheersing* 35, no. 2 (2013): 183–202.

*Burri, Monika, Kilian T. Elsasser, and David Gurgerli (eds.). *Die Internationalit der Eisenbahn 1850–1970* (Zürich: Chronos, 2003) (Interferenzen; Studien zur Kulturgeschichte der Technik, ed. David Gugerli, Vol. 7).

Butt, Joshua Richard. "Aspects of the Automobile's Diffusion in the North-West of England 1896–1939," unpublished dissertation (Manchester: Manchester Metropolitan University, 2018).

Butz, David, and Nancy Cook. "Mobile Methods, Epistemic Justice and Mobility Justice," in Nancy Cook and David Butz (eds.), *Mobilities, Mobility Justice and Social Justice* (London: Routledge, 2019), 81–97.

*Buzard, James. *The Beaten Track: European Tourism, Literature, and the Ways to Culture, 1800–1918* (Oxford: Clarendon Press, 1993).

Callaci, Emily. "Street Textuality: Socialism, Masculinity, and Urban Belonging in Tanzania's Pulp Fiction Publishing Industry, 1975–1985," *Comparative Studies in Society and History* 59, no. 1 (2017): 183–210.

*Callon, Michel. "The Sociology of an Actor-Network: The Case of the Electric Vehicle," in Michel Callon, John Law, and Arie Rip (eds.), *Mapping the Dynamics of Science and Technology: Sociology of Science in the Real World* (London: MacMillan Press, 1986), 19–34.

——. "The State and Technical Innovation: A Case Study of the Electrical Vehicle in France," *Research Policy* 9 (1980): 358–76.

Campanella, Thomas J. *The Concrete Dragon: China's Urban Revolution and What It Means for the World* (New York: Princeton Architectural Press, 2008).

*Carey, Mark. "Latin American Environmental History: Current Trends, Interdisciplinary Insights, and Future Directions," *Environmental History* 14 (April 2009): 221–52.

Carrier, Neil. "'*Miraa* is Cool': The Cultural Importance of *Miraa* (Khat) for Tigania and Igembe Youth in Kenya," *Journal of African Cultural Studies* 17, no. 2 (December 2015): 201–18.

Carter, G. F. "Automobiles for Average Incomes," *Outing* 55, no. 4 (January 1910): 410–19.

*Casey, Roger N. *Textual Vehicles: The Automobile in American Literature* (New York: Garland, 1997).

Cass, Noel, and Katharina Manderscheid. "The Automobility System: Mobility Justice and Freedom under Sustainability," in Nancy Cook and David Butz (eds.), *Mobilities, Mobility Justice and Social Justice* (London: Routledge, 2019), 101–15.

Cassamagnaghi, Silvia. "Vespa Piaggio: First Mass Motorization and Discovery of Leisure in Post-War Italy" (paper presented at the conference "Inventing Europe: Eurocores. European Science Foundation workshop, Amsterdam, 15–17 January 2009).

Cervero, Robert. *Informal Transport in the Developing World* (Nairobi: United Nations Centre for Human Settlements [Habitat], 2000).

Chang, Leslie T. *Factory Girls: From Village to City in a Changing China* (New York: Spiegel & Grau, 2008).

*Chatty, Dawn. *From Camel to Truck: The Bedouin in the Modern World* (Cambridge: White Horse Press, 2013 [1986]).

Chaturvedi, Vinayak. "Introduction," in Vinayak Chaturvedi (ed.), *Mapping Subaltern Studies and the Postcolonial* (London: Verso, 2000), vii–xix.

Chen, Eva. "Its Beauty, Danger, and Feverish Thrill: Speed and Cycling Women in Fin de Siècle Fiction," *MFS Modern Fiction Studies* 63, no. 4 (Winter 2017): 607–27.

——. "Fashioning the Cosmopolitan Girl: Sartorial Display and Technologies of Femininity in the Chinese Bestseller *Du Lala* (2007–11)," *Fashion Theory: The Journal of Dress, Body and Culture* 17, no. 5 (2013): 553–77.

——. "Freedom or Constraint: The Speeding Bicycle and the Industrial Subject in H.G. Wells's *The Wheels of Chance*," *Victorian Literature and Culture* 45 (2017): 433–47.

——. "'The Hate that Changed': Cycling Romance and the Aestheticization of Women Cyclists," *Victorian Periodicals Review* 53, no. 3 (Fall 2019): 489–517.

——. "Pretty Women Don't Drive: Early Women Motorists and the Taming of the Motor Car," *Women's Studies: An Interdisciplinary Journal* 49, no. 2 (2020): 149–79.

——. "Its Prohibitive Costs: The Bicycle, the New Woman and Conspicuous Display," *Journal of Language, Literature and Culture* 64, no. 1 (2017): 1–17.

——. "Shanghai(ed) Babies: Geopolitics, Biopolitics and the Global Chick Lit," *Feminist Media Studies* 12, no. 2 (2012): 214–28.

Chen, Eva Yin-I. "*Shanghai Baby* as a Chinese Chick-Lit: Female Empowerment and Neoliberal Consumerist Agency," *Asian Journal of Women's Studies* 15, no. 1 (2009): 54–93.

*Childs, William R. *Trucking and the Public Interest: The Emergence of Federal Regulation 1914–1940* (Knoxville: University of Tennessee Press, 1985).

Chilson, Peter. *Riding the Demon: On the Road in West Africa* (Athens: University of Georgia Press, 1999).

*Chilundo, Arlindo Gonçalo. "The Economic and Social Impact of Rail and Road Transportation Systems in the Colonial District of Moçambique (1900–1961)," unpublished dissertation (Minneapolis: University of Minnesota, November 1995).

Choi, Yoonjoung. "The Bi-Cycling Mr Hoopdriver: Counter-Sporting Victorian Reviving the Carnivalesque," *Critical Survey* 24, no. 1 (2012): 102–15.

*Clarke, Deborah. *Driving Women: Fiction and Automobile Culture in Twentieth-Century America* (Baltimore: Johns Hopkins University Press, 2007).

Clarsen, Georgine. "*Black As*: Performing Indigenous Difference," in Nancy Cook and David Butz (eds.), *Mobilities, Mobility Justice and Social Justice* (London: Routledge, 2019), 159–72.

*——. *Eat My Dust: Early Women Motorists* (Baltimore: Johns Hopkins University Press, 2008).

——. "Machines as the Measure of Women: Colonial Irony in a Cape to Cairo Automobile Journey, 1930," *Journal of Transport History* 29, no. 1 (March 2008): 44–63.

*——. "Pedaling Power: Bicycles, Subjectivities and Landscapes in a Settler Colonial Society," *Mobilities* 10, no. 5 (2015): 706–25.

*Cohan, Steven, and Ina Rae Hark (eds.). *The Road Movie Book* (London: Routledge, 1997).

Clift, Rebecca. "Irony in Conversation," *Language in Society* 28 (1999): 523–53.

Comaroff, Jean, and John L. Comaroff. *Theory from the South or, How Euro-America is Evolving toward Africa* (London: Routledge, 2012).

Connell, R. W. *Masculinities* (Cambridge: Polity Press, 1995).

——. *Southern Theory: The Global Dynamics of Knowledge in Social Science* (Cambridge: Polity Press, 2007).

Conrad, Sebastian. *What is Global History?* (Princeton, NJ: Princeton University Press, 2016).

Conrad, Sebastian, and Andreas Eckert. "Globalgeschichte, Globalisierung, multiple Modernen: Zur Geschichtsschreibung der modernen Welt," in Sebastian Conrad, Andreas Eckert, and Ulrike Freitag (eds.), *Globalgeschichte: Theorien, Ansätze, Themen* (Frankfurt: Campus, 2007), 7–49.

Conrad, Sebastian, and Klaus Mühlhahn. "Globale Mobilität und Nationalismus: Chinesische Migration und die Re-Territorialisierung des Nationalen um 1900," in Birgit Schäbler (ed.), *Area Studies und die Welt: Weltregionen und neue Globalgeschichte* (Vienna: Mandelbaum, 2007), 217–51.

Cowley, Malcolm. *Exile's Return: A Literary Odyssey of the 1920s*, ed. and with an introduction by Donald W. Faulkner (New York: Penguin, 1994 [1934]).

Cresswell, Tim. "Mobility as Resistance: A Geographical Reading of Kerouac's 'On the Road,'" *Transactions of the Institute of British Geographers* 18 (1993): 249–63.

——. *On the Move: Mobility in the Modern Western World* (New York: Routledge, 2006).

Crews, Harry. "The Car," in *Classic Crews: A Harry Crews Reader* (New York: Touchstone/Simon & Schuster, 1995), 323–28.

* Crosby, Alfred W. "Ecological Imperialism: The Overseas Migration of Western Europeans as a Biological Phenomenon," in Donald Worster (ed.), *The Ends of the Earth: Perspectives on Modern Environmental History* (Cambridge: Cambridge University Press, 1988) (Studies in Environment and History, eds. Donald Woster and Alfred Crosby).

Crouch, Martin. "Problems of Soviet Urban Transport," *Soviet Studies* 31, no. 2 (April 1979): 231–56.

Daggett, Cara. "Petro-Masculinity: Fossil Fuels and Authoritarian Desire," *Millennium: Journal of International Studies* 47, no. 1 (2018): 25–44.

* Dalakoglou, Dimitris, and Penny Harvey (eds.). *Roads and Anthropology: Ethnography, Infrastructures, (Im)mobility* (London: Routledge, 2015).

Dant, Tim, and Peter J. Martin. "By Car: Carrying Modern Society," in Jukka Gronow and Alan Warde (eds.), *Ordinary Consumption* (London: Routledge, 2001), 143–57.

* Davis, Clarence B., and Kenneth E. Wilburn Jr. (with Ronald E. Robinson) (eds.). *Railway Imperialism* (New York: Greenwood Press, 1991) (Contributions in Comparative Colonial Studies, No. 26).

* Davis, Donald F. "Competition's Moment: The Jitney-Bus and Corporate Capitalism in the Canadian City, 1914–29," *Urban History Review/Revue d'histoire urbaine* 18, no. 2 (October 1989): 101–22.

* Davison, Graeme (with Sheryl Yelland). *Car Wars: How the Car Won Our Hearts and Conquered Our Cities* (Crows Nest: Allen & Unwin, 2004).

De Jong, Wim, and Bas van Kleef. *De Puch-story en andere brommerverhalen* (Bussum: Thoth, 1994).

Denning, Andrew. *Automotive Empire: How Cars and Roads Fueled European Colonialism in Africa* (Ithaca, NY: Cornell University Press, 2024).

Dennis, Kingsley, and John Urry. *After the Car* (Cambridge: Polity Press, 2009).

Desportes, Marc. *Paysages en mouvement: Transports et perception de l'espace XVIIIe–XXe siècle* (n.p. [Paris]: Gallimard, 2005).

* Devauges, Jean-Denys, and Didier Masseau (eds.). *Le voyage en France 1750–1914: Colloque de Compiègne 30 et 31 Mai 1997* (n.p., n.d).

* Dick, Howard, and Peter J. Rimmer. *Cities, Transport and Communications: The Integration of Southeast Asia since 1850* (New York: Palgrave Macmillan, 2003).

* Dickson, K. B. "The Development of Road Transport in Southern Ghana and Ashanti since about 1850," *Transactions of the Historical Society of Ghana* 5, no. 1 (1961): 33–42.

Dikötter, Frank. *Exotic Commodities: Modern Objects and Everyday Life in China* (New York: Columbia University Press, 2006).

Dimitriou, Harry T. "A Developmental Approach to Urban Transport Planning: The Urban Transport Planning Process and Its Derivatives: An Assessment of their Contribution to Transport Planning for Third World Cities; Volume I," doctoral dissertation (Cardiff: University of Wales College of Cardiff, Department of Town Planning, October 1989).

*Disko, Sasha. *The Devil's Wheels: Men and Motorcycling in the Weimar Republic* (Oxford: Berghahn, 2016).

———. "'The World is My Domain'—Technology, Gender and Orientalism in German Interwar Motorized Adventure Literature," *Transfers* 1, no. 3 (Fall 2011): 44–63.

Ditz, Toby L. "The New Men's History and the Peculiar Absence of Gendered Power: Some Remedies from Early American Gender History," *Gender & History* 16, no. 1 (April 2004): 1–35.

*Divall, Colin, and George Revill. "Cultures of Transport: Representation, Practice and Technology," *Journal of Transport History* 26, no. 1 (March 2005): 99–111.

Dobson, Wendy. *Living with China: A Middle Power Finds Its Way* (Toronto: University of Toronto Press, 2019).

Doel, Marcus A. "Representation and Difference," in Ben Anderson and Paul Harrison (eds.), *Taking-Place: Non-Representational Theories and Geography* (Farnham: Ashgate, 2010), 117–30.

Domonoske, Camila. "Giving up Gas-Powered Cars Was a Fringe Idea. It's Now on Its Way to Reality," *NPR*, 20 November 2021. Retrieved 20 March 2024 from https://www.npr.org/2021/11/20/1055718914/.

Drayton, Richard, and David Motadel. "Discussion: The Futures of Global History," *Journal of Global History* 13 (2008): 1–21.

Duarte, Fábio, and Carlo Ratti. "The Impact of Autonomous Vehicles on Cities: A Review," *Journal of Urban Technology* 25, no. 4 (2018): 3–18.

*Duc, Gérard, Olivier Perroux, Hans-Ulrich Schiedt, and François Walter (eds.). *Histoire des transports et de la mobilité: Entre concurrence modale et coordination (de 1918 à nos jours)/Transport and Mobility History: Between Modal Competition and Coordination (from 1918 to the Present)* (Neuchâtel: Alphil, 2014).

*Dyos, H. J., and D. H. Aldcroft. *British Transport: An Economic Survey from the Seventeenth Century to the Twentieth* (Harmondsworth: Penguin, 1969).

*Ebert, Anne-Katrin. "Cycling towards the Nation: The Use of the Bicycle in Germany and the Netherlands, 1880–1940," *European Review of History* 11, no. 3 (2004): 347–64.

*———. "Ein Ding der Nation? Das Fahrrad in Deutschland und den Niederlanden, 1880–1940. Eine vergleichende Konsumgeschichte," doctoral dissertation (Bielefeld: Universität Bielefeld, Fakultät für Geschichtswissenschaft, Philosophie und Theologie, January 2009).

Edensor, Tim, and Mark Jayne. "Introduction: Urban Theory beyond the West," in Tim Edensor and Mark Jayne (eds.), *Urban Theory beyond the West: A World of Cities* (London: Routledge, 2012), 1–27.

Edgerton, David. "Creole Technologies and Global Histories: Rethinking How Things Travel in Space and Time," *HOST: Journal of History of Science and Technology* 1 (Summer 2007): 75–112.

Ehrenburg, Ilja [Ilya]. *Das Leben der Autos*, trans. Hans Ruoff (Amsterdam: Paco-Press, 1973 [1929]).

*Elliot, Lyn Elizabeth. "Liberating Domesticity: The American Road Narrative in Film and Fiction," unpublished dissertation (Iowa City: University of Iowa, December 2000).

*Emanuel, Martin, Frank Schipper, and Ruth Oldenziel (eds.). *A U-Turn to the Future: Sustainable Urban Mobility since 1850* (New York: Berghahn, 2020).

Emgin, Bahar. "Rowdy Gangs on the Road: The Culture of Mobilet in Turkey," in Tevfik Balcioğlu and Gülsüm Baydar (eds.), *Nesneyi Okumak/Deciphering the Object* (Izmir: Yaşar Üniversitesi, 2012), 36–46.

"The End of the Road? An Overview of Combustion-Engine Car Phase-out Announcements across Europe" (Briefing International Council on Clean Transportation, ICCT, May 2020). Retrieved 20 March 2024 from https://theicct.org/sites/default/files/publications/.

*Endy, Christopher. *Cold War Holidays: American Tourism in France* (Chapel Hill: University of North Carolina Press, 2004) (The New Cold War History, ed. John Lewis Gladdis).

Esty, Jed, and Colleen Lye. "Peripheral Realisms Now," *Modern Language Quarterly* 73, no. 3 (September 2012): 269–88.

Engerman, David C. "Introduction: Histories of the Future and the Futures of History," *American Historical Review* 117 (December 2012): 1402–10.

Escobar, Arturo. *Encountering Development: The Making and Unmaking of the Third World* (Princeton, NJ: Princeton University Press, 2012).

*Eyerman, Ron, and Orvar Löfgren. "Romancing the Road: Road Movies and Images of Mobility," *Theory, Culture & Society: Explorations in Critical Social Science* 12, no. 1 (1995): 53–79.

Fagnant, Daniel J., and Kara Kockelman. "Preparing a Nation for Autonomous Vehicles: Opportunities, Barriers and Policy Recommendations," *Transportation Research Part A* 77 (2015): 167–81.

*Fein, Michael R. *Paving the Way: New York Road Building and the American State, 1880–1956* (Lawrence: University Press of Kansas, 2008).

Fell, Karolina Dorothea. *Kalkuliertes Abenteuer: Reiseberichte deutschsprachiger Frauen (1920–1945)* (Stuttgart: J. B. Metzler, 1998).

*Filarski, Ruud. *The Rise and Decline of Transport Systems: Changes in a Historical Context* (Rotterdam: Ministry of Transport and Public Works, AVV Transport Research Centre, 2004).

*Filarski, Ruud (in cooperation with Gijs Mom). *Shaping Transport Policy; Two centuries of struggle between the public and private sector – A comparative perspective* (The Hague: Sdu Uitgevers, 2011).

Filarski, Ruud, and Gijs Mom. *Geschiedenis van de verkeersveiligheid in Nederland* [History of traffic safety in the Netherlands] (forthcoming 2025).

*Filarski, Ruud, and Gijs Mom. *Van transport naar mobiliteit: De transportrevolutie (1800–1900)* (Zutphen: Walburg Pers, 2008).

*Flik, Reiner. *Von Ford lernen? Automobilbau und Motorisierung in Deutschland bis 1933* (Cologne: Böhlau, 2001) (Wirtschafts- und Sozialhistorische Studien, ed. Stuart Jenks, Michael North, and Rolf Walter, Vol. 11).

*Flink, James J. *America Adopts the Automobile, 1895–1910* (Cambridge, MA: MIT Press, 1970).

*——. *The Automobile Age* (Cambridge, MA: MIT Press, 1993 [1988]).

*Flonneau, Mathieu. *L'Automobile au temps des Trentes Glorieuses: Un rêve d'automobilisme* (Carbonne: Nouvelle Éditions Loubatières, 2016).

*——. *L'Autorefoulement et ses limites: Raisonner l'impensable mort de l'automobile* (Paris: Descartes & Cie, 2010) (Collection "Cultures mobiles," eds. Mathieu Flonneau and Arnaud Passalacqua).

*——. "Une histoire des taxis parisiens au cours de la première moitié du XXe siècle: Entre innovation, occasions manquées et blocages," in *Paris et Ile-de-France: Mémoires publiés par la Fédération des Sociétés Historiques et Archéologiques de Paris et de L'Ile-de-France, Tome 62* (Paris: Klincksieck, 2011), 37–53.

*——. *Paris et l'automobile: Un siècle de passions* (Paris: Hachette Littératures, 2005).

*Fogel, Joshua A. *The Literature of Travel in the Japanese Rediscovery of China 1862–1945* (Stanford, CA: Stanford University Press, 1996).

Foreman-Peck, James, Sue Bowden, and Alan McKinley. *The British Motor Industry* (Manchester: Manchester University Press, 1995) (British Industries in the Twentieth Century, ed. Derek H. Aldcroft).

*Franz, Kathleen. *Tinkering: Consumers Reinvent the Early Automobile* (Philadelphia: University of Pennsylvania Press, 2005).

*Fraunholz, Uwe. "'Automobilistisches Feindesland': Motorisierte Schweiz-Touristen in der Pionierzeit des Autos," in Bruno Meier (ed.), *Autolust: Ein Buch über die Emotionen des Autofahrens* (Baden: hier + jetzt, 2002), 120–25.

*——. *Motorphobia: Anti-automobiler Protest in Kaiserreich und Weimarer Republik* (Göttingen: Vandenhoeck & Ruprecht, 2002).

Freedman, Alisa. *Tokyo in Transit: Japanese Culture on the Rails and Road* (Stanford, CA: Stanford University Press, 2011).

Freeman, John Brian. "Transnational Mechanics: Automobility in Mexico, 1895–1950," unpublished dissertation (New York: City University of New York, 2012).

Frello, Birgitta. "Towards a Discursive Analytics of Movement: On the Making and Unmaking of Movement as an Object of Knowledge," *Mobilities* 3, no. 1 (2008): 25–50.

Freund, Peter, and George Martin. "Driving South: The Globalization of Auto Consumption and Its Social Organization of Space," *Capitalism, Nature, Socialism* 11, no. 4 (December 2000): 51–71.

Frith, Simon. *Sound Effects: Youth, Leisure, and the Politics of Rock 'n' Roll* (New York: Pantheon, 1981).

*Fulwider, Benjamin. "Driving the Nation: Road Transportation and the Postrevolutionary Mexican State, 1925–1960," unpublished dissertation (Washington, DC: Georgetown University, 2009).

*Furlough, Ellen. "Une leçon des choses: Tourism, Empire, and the Nation in Interwar France," *French Historical Studies* 25, no. 3 (Summer 2002): 441–73.

*——. "Making Mass Vacations: Tourism and Consumer Culture in France, 1930s to 1970s," *Comparative Studies in Society and History* 40 (1998): 247–86.

*Fussell, Paul. *Abroad: British Literary Traveling between the Wars* (Oxford: Oxford University Press, 1980).

Gagica, Rita, and Ger Duijzings. "Liberated or Lawless? Social Life on Prishtina's Postwar Streets," in Ger Duijzings and Tauri Tuvikene (eds.), *If Cars Could Walk: Postsocialist Streets in Transformation* (New York: Berghahn, 2023), 46–69 (Explorations in Mobility, eds. Georgine Clarsen et al.).

Gänger, Stefanie. "Circulation: Reflections on Circularity, Entity, and Liquidity in the Language of Global History," *Journal of Global History* 12 (2017): 303–18.

*Gartman, David. *Auto Opium: A Social History of American Automobile Design* (New York: Routledge, 1994).

——. "Three Ages of the Automobile: The Cultural Logics of the Car," *Theory, Culture & Society* 21, no. 4/5 (2004): 169–95.

Gatejel, Luminita. "The Common Heritage of the Socialist Car Culture," in Lewis H. Siegelbaum (ed.), *The Socialist Car: Automobility in the Eastern Bloc* (Ithaca, NY: Cornell University Press, 2011), 143–56.

——. "Privat oder Staatlich? Automobile Konsumkultur in der Sowjetunion, der DDR und Rumänien," in Manuel Schramm (ed.), *Vergleich und Transfer in der Konsumgeschichte* (Leipzig: Leipziger Universitätsverlag, 2009), 16–32.

——. *Warten, hoffen und endlich fahren: Auto und Sozialismus in der Sowjetunion, in Rumänien und der DDR (1956–1989/91)* (Frankfurt: Campus, 2014).

——. "The Wheels of Desire; Automobility Discourses in the Soviet Union," in Manfred Krieger, Ulrike Gutzmann, and Dirk Schlinkert (eds.), *Towards Mobility: Varieties of Automobilism in East and West* (Wolfsburg: Volkswagen AG, 2009), 31–41 (Schriften zur Unternehmensgeschichte von Volkswagen, 3).

Gebhardt, Laura, Christian Wolf, and Robert Seiffert. "'I'll Take the E-Scooter Instead of My Car'—The Potential of E-Scooters as a Substitute for Car Trips in Germany," *Sustainability* 13 (2021): 7361.

Geels, F. W. "Processes and Patterns in Transitions and System Innovations: Refining the Co-Evolutionary Multi-Level Perspective," *Technological Forecasting and Social Change* 72 (2005): 681–96.

——. "The Multi-Level Perspective on Sustainable Transitions: Responses to Seven Criticisms," *Environmental Innovation and Societal Transitions* 1 (2011): 24–40.

——. "A Socio-Technical Analysis of Low-Carbon Transitions: Introducing the Multi-Level Perspective into Transport Studies," *Journal of Transport Geography* 24 (2012): 471–82.

Geels, F. W., Geoff Dudley, and René Kemp. "Findings, Conclusions and Assessments of Sustainability Transitions in Automobility," in Frank W. Geels, René Kemp, Geoff Dudley, and Glenn Lyons (eds.), *Automobility in Transition? A Socio-Technical Analysis of Sustainable Transport* (New York: Routledge, 2012), 335–73.

Genus, Audley, and Anne-Marie Coles. "Rethinking the Multi-Level Perspective of Technological Transitions," *Research Policy* 37 (2008): 1436–45.

*Gerhold, Dorian (ed.). *Road Transport in the Horse-Drawn Era* (Aldershot: Ashgate, 1966) (Studies in Transport History, ed. John Armstrong).

Gerland, Ellen J., and Craig L. Davis. "Selling Detroit on Women: *Woman's Day* and Auto Advertising, 1964–82," *Journalism History* 38, no. 4 (Winter 2013): 209–20.

*Gewald, Jan-Bart, Sabine Luning, and Klaas van Walraven (eds.). *The Speed of Change: Motor Vehicles and People in Africa, 1890–2000* (Leiden: Brill, 2009).

Ghobrial, John-Paul A. "Introduction: Seeing the World like a Microhistorian," *Past and Present* (2019), Supplement 14: 1–22.

Gibson, James J. "The Theory of Affordances," in Robert Shaw and John Bransford (eds.), *Perceiving, Acting, and Knowing: Toward an Ecological Psychology* (Hillsdale, NJ: Lawrence Erlbaum, 1977), 67–82.

*Gladstone, David L. *From Pilgrimage to Package Tour: Travel and Tourism in the Third World* (New York: Routledge, 2005).

*Glaister, Stephen, and Corinne Mulley. *Public Control of the British Bus Industry* (Aldershot: Gower, 1983).

*Gleber, Anke. *The Art of Taking a Walk: Flanerie, Literature, and Film in Weimar Culture* (Princeton, NJ: Princeton University Press, 1999).

*Goddard, Stephen B. *Getting There: The Epic Struggle between Road and Rail in the American Century* (New York: Basic Books, 1994).

Goodwin, Katherine J. "Reconstructing Automobility: The Making and Breaking of Modern Transportation," *Global Environmental Politics* 10, no. 4 (November 2010): 60–78.

*Gopakumar, Govind. *Installing Automobility: Emerging Politics of Mobility and Streets in Indian Cities* (Cambridge, MA: MIT Press, 2020).

*Gorsuch, Anne E. "'There's No Place Like Home': Soviet Tourism in Late Stalinism," *Slavic Review* 62, no. 4 (Winter 2003): 760–85.

Gössling, Stefan. *The Psychology of the Car: Automobile Admiration, Attachment, and Addiction* (Amsterdam: Elsevier, 2017).

Gould, Peter R. *The Development of the Transportation Pattern in Ghana* (Evanston, IL: Department of Geography, Northwestern University, April 1960).

*Grace, Joshua. *African Motors: Technology, Gender, and the History of Development* (Durham, NC: Duke University Press, 2021).

——. "Heroes of the Road: Race, Gender and the Politics of Mobility in Twentieth Century Tanzania," *Africa* 83, no. 3 (2013): 403–25.

Grivel, Charles. "Automobil: Zur Genealogie subjektivischer Maschinen," in Martin Stingelin and Wolfgang Scherer (eds.), *HardWar/SoftWar: Krieg und Medien 1914 bis 1945* (Munich: Wilhelm Fink, 1991), 171–96.

*——. "Automobile et cinéma I," in Michel Bouvier, Michel Lartouce, and Lucie Roy (eds.), *Cinéma: Acte et présence* (Quebec: Éditions Nota Bene, 1999), 63–85.

*——. "D'un écran automobile," in Jochen Mecke and Volker Roloff (eds.), *Kino-/(Ro) Mania: Intermedialität zwischen Film und Literatur* (Tübingen: Stauffenburg, 1999) (Siegener Forschungen zur romanischen Literatur- und Medienwissenschaft, ed. Volker Roloff, 1), 47–77.

Gronow, Jukka, and Sergei Zhuravlev. "Soviet Luxuries from Champagne to Private Cars," in David Crowley and Susan E. Reid (eds.), *Pleasures in Socialism: Leisure and Luxury in the Eastern Bloc* (Evanston, IL: Northwestern University Press, 2010), 121–46.

*Gross, Andrew. "California Automobile Tourism and Consumer Culture in American Literature, 1916 to 1939" (Frankfurt am Main: Johann Wolfgang Goethe-Universität, Zentrum für Nordamerika-Forschung, ZENAF Arbeits- und Forschungsberichte, No. 1, 2003).

*Grübler, Arnulf. *The Rise and Fall of Infrastructures: Dynamics of Evolution and Technological Change in Transport* (Heidelberg: Physica, 1990).

Gualthérie van Weezel, Tjerk. "Hoe 197 landen het eens werden over het einde van fossiele energie" [How 197 countries agreed on the end of fossil energy], *Volkskrant* (14 December 2023): 13–15.

Hahn, Hans Peter. "The Appropriation of Bicycles in West Africa: Pragmatic Approaches to Sustainability," *Transfers* 2, no. 2 (Summer 2012): 31–48.

Haldrup, Michael. "What Futures? Our Future, the Planet, and Cohabitation," *Transfers* 13, no. 1/2 (Spring/Summer 2023): 144–61.

*Hamilton, Shane L. *Trucking Country; The Road to America's Wal-Mart Economy* (Princeton/Oxford: Princeton University Press, 2008) (Politics and Society in Twentieth-Century America, eds.: William Chafe, Gary Gerstle, Linda Gordon and Julian Zelizer).

Hamilton-Baillie, Ben. "Shared Space: Reconciling People, Places and Traffic," *Built Environment* 34, no. 2 (2008): 161–81.

Hansen, Arve. "*Doi moi* on Two and Four Wheels: Capitalist Development and Motorised Mobility in Vietnam," in Arve Hansen and Kenneth Bo Nielsen (eds.), *Cars, Automobility and Development in Asia: Wheels of Change* (London: Routledge, 2017), 103–19.

Hanson, Susan. "Transportation: Hooked on Speed, Eyeing Sustainability," in Eric Sheppard and Trevor K. J. Barnes (eds.), *A Companion to Economic Geography* (Oxford: Blackwell, 2000) (Blackwell Companions to Geography), 468–83.

Harmsworth, Alfred C. *Motors and Motor-Driving* (London: Longmans Green, 1902).

Harrell, Stevan. *An Ecological History of Modern China* (Seattle: University of Washington Press, 2023).

*Hård, Mikael (ed.). *Automobile Engineering in a Dead End: Mainstream and Alternative Developments in the 20th Century* (Gothenburg: Gothenburg University, 1992).

*Hård, Mikael, and A. Jamison. "Alternative Cars: The Contrasting Stories of Steam and Diesel Automotive Engines," *Technology in Society* 19, no. 2 (1997): 145–60.

*Hård, Mikael, and Thomas J. Misa. *The Urban Machine: Recent Literature on European Cities in the 20th Century* (Cambridge, MA: MIT Press, 2008).

Hauser, Heinrich. *Fahrten und Abenteuer im Wohnwagen: Herausgegeben, kommentiert und mit einem Nachwort versehen von Robert Hilgers* (Stuttgart: Dolde-Medien, 2004 [1935]).

——. *Schwarzes Revier, Herausgegeben von Barbara Weidle: Mit einem Vorwort von Andreas Rossmann* (Bonn: Weidle, 2010 [1930]).

*Hawkins, E. K. *Roads and Road Transport in an Underdeveloped Country: A Case Study of Uganda* (London: Her Majesty's Stationery Office, 1962) (Colonial Office, Colonial Research Studies No. 32).

Headrick, Daniel R. *The Tools of Empire: Technology and European Imperialism in the Nineteenth Century* (New York: Oxford University Press, 1981).

Hebdige, Dick. *Hiding in the Light: On Images and Things* (London: Routledge, 1988).

Heilweil, Rebecca. "It's the End of the Car as We Know It: The Electric Vehicle Transition is Really a U-Turn," *Vox*, 2 September 2022. Retrieved 20 March 2024 from https://www.vox.com/recode/23333356/ev-electric-vehicle-self-driving-cars-internal-combustion-driving.

Heinrichs, Dirk, Daniel Ehebrecht, and Barbara Lenz. "Moving beyond Informality? Theory and Reality of Public Transport in Urban Africa," in Tanu Priya Uteng and Karen Lucas (eds.), *Urban Mobilities in the Global South* (London: Routledge, 2017), 134–54.

Heinze, Robert. "Fighting over Urban Space: Matatu Infrastructure and Bus Stations in Nairobi, 1960–2000," *Africa Today* 65, no. 2 (Winter 2018): 2–21.

——. "'Taxi Pirates': A Comparative History of Informal Transport in Nairobi and Kinshasa, 1960s–2000s," in Daniel E. Agbiboa (ed.), *Transport, Transgression and Politics in African Cities: The Rhythm of Chaos* (Abingdon: Routledge, 2019) (Cities and Society, ed. Chris Pickvance), 19–39.

Heller, Dana A. *The Feminization of Quest-Romance: Radical Departures* (Austin: University of Texas Press, 1990).

Henderson, Jason. "EVs Are Not the Answer: A Mobility Justice Critique of Electric Vehicle Transitions," *Annals of the American Association of Geographers* 110, no. 6 (2020): 1993–2019.

——. "The Politics of Mobility: De-Essentializing Automobility and Contesting Urban Space," in Jim Conley and Arlene Tigar McLaren (eds.), *Car Troubles: Critical Studies of Automobility and Auto-Mobility* (Farnham: Ashgate, 2009), 147–64.

——. "Secessionist Automobility: Racism, Anti-Urbanism, and the Politics of Automobility in Atlanta, Georgia," *International Journal of Urban and Regional Research* 30, no. 2 (June 2006): 293–307.

*Héran, Frédéric. *Le retour de la bicyclette: Une histoire des déplacements urbains en Europe, de 1817 à 2050* (Paris: La Découverte, 2014).

Hertling, Anke. *Eroberung der Männerdomäne Automobil: Die Selbstfahrerinnen Ruth Landshoff-Yorck, Erika Mann und Annemarie Schwarzenbach* (Bielefeld: Aisthesis, 2013).

Herwig, Holger H. *The Marne, 1914: The Opening of World War I and the Battle That Changed the World* (New York: Random House, 2009).

*Heslop, Luke, and Galen Murton (eds.). *Highways and Hierarchies: Ethnographies of Mobility from the Himalaya to the Indian Ocean* (Amsterdam: Amsterdam University Press, 2021) (New Mobilities in Asia, ed. Pál Myíri).

*Hey, Kenneth. "Cars and Films in American Culture, 1929–1959," *Michigan Quarterly Review* 19, no. 4 (1980): 588–600.

*Heymel, Charlotte. *Touristen an der Front: Das Kriegserlebnis 1914–1918 als Reiseerfahrung in zeitgenössischen Reiseberichten* (Berlin: LIT, 2007) (Literatur—Kultur—Medien, ed. Peter J. Brenner, 7).

Higashide, Steven. "Never Again Is Now: The Transportation Professions' Responsibility to Work Toward Justice," in Ryan Reft, Amanda K. Phillips de Lucas, and Rebecca C. Retzlaff (eds.), *Justice and the Interstates: The Racist Truth about Urban Highways* (Washington, DC: Island Press, 2023), 175–92.

Highmore, Ben. *Cityscapes: Cultural Readings in the Material and Symbolic City* (New York: Palgrave Macmillan, 2005).

Hill, Nikolas, et al. "Determining the Environmental Impacts of Conventional and Alternatively Fuelled Vehicles through LCA: Final Report for the European Commission" (Brussels: European Commission, Directorate-General for Climate Policy, Directorate C—Climate Strategy, Governance and Emissions from Non-Trading Sectors, Unit C4—Road Transport, 13 July 2020).

*Hodges, Graham Russell Gao. *Taxi! A Social History of the New York City Cabdriver* (Baltimore: Johns Hopkins University Press, 2007).

Hoerder, Dirk. *Cultures in Contact: World Migration in the Second Millennium* (Durham, NC: Duke University Press, 2002).

Hoffmann, Lukas. *Postirony: The Nonfiction Literature of David Foster Wallace and Dave Eggers* (Bielefeld: transcript, 2016).

*Hofmeier, Rolf. *Transport and Economic Development in Tanzania, with Particular Reference to Roads and Road Transport* (Munich: Weltforum, 1973) (Afrika-Studien, ed. Wilhelm Marquardt and Hans Ruthenberg; Economic Research Bureau, University College, Dar es Salaam/IFO-Institut für Wirtschaftsforschung München, Afrika-Studienstelle).

*Horner, Craig. *The Emergence of Bicycling and Automobility in Britain* (London: Bloomsbury Academic, 2021).

Howe, John. "Sustaining Africa's Rural Networks: The Asset Management Approach," *World Transport Policy & Practice* 5 (1999): 11–16.

*Hugill, Peter J. "The Rediscovery of America: Elite Automobile Touring," *Annals of Tourism Research* 12, no. 3 (1985): 435–47.

Hunter, Holland. *Soviet Transport Experience: Its Lessons for Other Countries* (Washington, DC: Brookings Institution, Transport Research Program, 1968).

Hussain, Yasmin. *Writing Diaspora: South Asian Women, Culture and Ethnicity* (Aldershot: Ashgate, 2005).

Infrastructure at the Crossroads: Lessons from 20 Years of World Bank Experience (Washington, DC: World Bank, 2006).

Ingold, Tim. "When ANT Meets SPIDER: Social Theory for Arthropods," in C. Knappett and L. Malafouris (eds.), *Material Agency: Towards a Non-Anthropocentric Approach* (New York: Springer, 2008), 209–15.

Inkinen, Tommi, Tan Yigitcanlar, and Mark Wilson. "Smart Cities and Innovative Urban Technologies," *Journal of Urban Technology* 26, no. 2 (2019): 1–2.

Isenberg, Nancy. *White Trash: The 400-Year Untold History of Class in America* (New York: Viking, 2016).

Jacobsen, P. L. "Safety in Numbers: More Walkers and Bicyclists, Safer Walking and Bicycling," *Injury Prevention* 9 (2003): 205–9.

*Jakle, John A. "Touring by Automobile in 1932: The American West as Stereotype," *Annals of Tourism Research* 8, no. 4 (1981): 534–49.

Janssens, Sander. "Kapot mooie auto" [Wrecking our lovely cars], *Volkskrant* (15 June 2024): V20.

Jastrząb, Mariusz. "Cars as Favors in People's Poland," in Lewis H. Siegelbaum, *The Socialist Car: Automobility in the Eastern Bloc* (Ithaca, NY: Cornell University Press, 2011), 30–46.

Jazeel, Tariq, and Stephen Legg. "Subaltern Studies, Space, and the Geographical Imagination," in Tariq Jazeel and Stephen Legg (eds.), *Subaltern Geographies* (Athens: University of Georgia Press, 2019) (Geographies of Justice and Social Transformation, eds. Mathew Coleman and Sapana Doshi), 1–35.

Kahn, Naveeda. *In Quest of a Shared Planet: Negotiating Climate from the Global South* (New York: Fordham University Press, 2023).

Kębłowski, Wojciech, and David Bassens. "'All Transport Problems Are Essentially Mathematical': The Uneven Resonance of Academic Transport and Mobility Knowledge in Brussels," *Urban Geography* 39, no. 3 (2018): 413–37.

Kębłowski, Wojciech, Frédéric Dobruszkes, and Kobe Boussauw. "Moving Past Sustainable Transport Studies: Towards a Critical Perspective on Urban Transport," *Transportation Research Part A: Policy and Practice* 159 (2022) [reflowed pagination; 9].

Keck-Szajbel, Mark. "Hitchikers' Paradise: The Intersection of Mass Mobility, Consumer Demand, and Ideology in the People's Republic of Poland," in Cathleen M. Giustino, Catherine J. Plum, and Alexander Vari (eds.), *Socialist Escapes: Breaking Away from Ideology and Everyday Routine in Eastern Europe, 1945–1989* (New York: Berghahn, 2013), 167–86.

Kennard, [Mary]. *The Motor Maniac: A Novel* (London: Hutchinson, 1902).

Kern, Stephen. *The Culture of Time and Space 1880–1918* (London: Harvard University Press, 1983).

Kesselring, Sven, and Gerlinde Vogl. "The New Mobilities Regime," in Susanne Witzgall, Gerlinde Vogl, and Sven Kesselring (eds.), *New Mobilities Regimes in Art and Social Sciences* (Farnham: Ashgate, 2013), 17–36.

Khairullina, Elvira, and Luis Santos y Ganges. "Tram, Trolleybus and Bus Services in Eastern-European Socialist Urban Planning: Case Studies of Magdeburg, Ostrava and Oryol (1950s and 1960s)," *Journal of Transport History* 42, no. 1 (2021): 26–57.

Kim, Han Sang. "'My Car' Modernity: What the U.S. Army Brought to South Korean Cinematic Imagination about Modern Mobility," *Journal of Asian Studies* 75, no. 1 (February 2016): 63–85.

Kim, Seung-Kuk. "Changing Lifestyles and Consumption Patterns of the South Korean Middle Class and New Generations," in Chua Beng-Huat (ed.), *Consumption in Asia: Lifestyles and Identities* (London: Routledge, 2000), 61–81.

*Kirsch, David A. *The Electric Vehicle and the Burden of History* (New Brunswick, NJ: Rutgers University Press, 2000).

Klopp, Jacqueline M., and Clemence M. Cavoli. "The Paratransit Puzzle: Mapping and Master Planning for Transportation in Maputo and Nairobi," in Tanu Priya Uteng and Karen Lucas (eds.), *Urban Mobilities in the Global South* (London: Routledge, 2017), 95–110.

*Knie, Andreas. *Diesel—Karriere einer Technik: Genese und Formierungsprozesse im Motorenbau* (Berlin: Sigma Bohn, 1991).

*——. *Wankel-Mut in der Autoindustrie: Anfang und Ende einer Antriebsalternative* (Berlin, 1994).

Knowles, Caroline. "Cities on the Move: Navigating Urban Life," *City* 15, no. 2 (2011): 135–53.

Kong, Shuyu. "Swan and Spider Eater in Problematic Memoirs of Cultural Revolution," *Positions* 7, no. 1 (1999): 239–52.

*Kopper, Christopher M. "The Breakthrough of the Package Tour in Germany after 1945," *Journal of Tourism History* 1, no. 1 (March 2009): 67–92.

*Kopper, Christopher, and Massimo Moraglio (eds.). *The Organization of Transport: A History of Users, Industry, and Public Policy* (London: Routledge, 2015).

Korstanje, Maximiliano E. *The Mobilities Paradox: A Critical Analysis* (Cheltenham: Edward Elgar, 2018).

*Koshar, Rudy (ed.). *Histories of Leisure* (Oxford: Berg, 2002) (Leisure, Consumption and Culture, ed. Rudy Koshar).

*Kramer, Dieter, and Ronald Lutz (eds.). *Reisen und Alltag: Beiträge zur kulturwissenschaftlichen Tourismusforschung* (Frankfurt: Institut für Kulturanthropologie und Europäische Ethnologie der Universität Frankfurt am Main, 1992) (Kulturanthropologie-Notizen, 39).

*Krämer-Bodoni, Thomas, Herbert Grymer, and Marianne Rodenstein. *Zur sozioökonomischen Bedeutung des Automobils* (Frankfurt am Main: Suhrkamp, 1971) (edition suhrkamp, 540, ed. Günther Busch).

*Krebs, Stefan. "'Notschrei eines Automobilisten' oder die Herausbildung des deutschen Kfz-Handwerks in der Zwischenkriegszeit," *Technikgeschichte* 79, no. 3 (2012): 185–206.

Kröger, Fabian. *From Automated to Autonomous Driving: A Transnational Research History on Pioneers, Artifacts and Technological Change (1950–2000)* (Cham: Springer, 2023) (Archimedes, New Studies in the History and Philosophy of Science and Technology, 70, ed. Jed Z. Buchwald).

Kröss, Verena. "Profitability through Quantification: The World Bank and Rural Development in the 1970s," in Corinne R. Unger, Nicholas Ferns, Jack Loveridge, and Iris Borowy (eds.), *Perspectives on the History of Global Development* (Berlin: Walter de Gruyter, 2022) (*Yearbook for the History of Global Development*, vol. 1, eds. Iris Borowy and Jack Loveridge), 183–204.

Kruglova, Anna. "Driving in Terrain: Automobility, Modernity, and the Politics of Statelessness in Russia," *American Ethnologist* 46, no. 4 (2019): 457–69.

Kuhr-Korolev, Corinna, and Dirk Schlinkert. "Introduction," in Manfred Grieger, Ulrike Gutzmann, and Dirk Schlinkert (eds.), *Towards Mobility: Varieties of Automobilism in East and West* (Wolfsburg: Volkswagen AG, 2009) (Schriften zur Unternehmensgeschichte von Volkswagen, 3), 9–20.

Ladd, Brian. *Autophobia: Love and Hate in the Automotive Age* (Chicago: University of Chicago Press, 2008).

*Laderman, David. *Driving Visions: Exploring the Road Movie* (Austin: University of Texas Press, 2002).

Lao, She. *Rickshaw: The Novel Lo-t'o Hsiang Tzu*, trans. Jean M. James (Honolulu: University Press of Hawaii, 1979).

Larsen, Jonas. "Tourism Mobilities and the Travel Glance: Experiences of Being on the Move," *Scandinavian Journal of Hospitality and Tourism* 1, no. 2 (2001): 80–98.

Laurier, Eric, et al. "Driving and 'Passengering': Notes on the Ordinary Organization of Car Travel," *Mobilities* 3, no. 1 (March 2008): 1–23.

*Laux, James M. *The European Automobile Industry* (New York: Twayne, 1992).

Law, John. "After ANT: Complexity, Naming and Topology," in John Law and John Hassard (eds.), *Actor Network Theory and After* (Oxford: Blackwell, 1999) (*Sociological Review* Monographs), 1–14.

*Lawrence, Karen R. *Penelope Voyages: Women and Travel in the British Literary Tradition* (Ithaca, NY: Cornell University Press, 1994).

Layborn, Keith, with David Taylor. *The Battle for the Roads of Britain: Police, Motorists and the Law, c.1890s to 1970s* (New York: Palgrave Macmillan, 2015).

Lee, Yong Sang. "Session Description" (Session "History of Korean Railroad: Geography, Geopolitics and Imaginative Geography," 2023 GMHC&T2M Annual Conference, Seoul, Konkuk University, 25–28 October 2023).

*Lekan, Thomas, and Thomas Zeller (eds.). *Germany's Nature: Cultural Landscapes and Environmental History* (New Brunswick, NJ: Rutgers University Press, 2005).

Lelièvre, Marie-Dominique. *Sagan à toute allure* (n.p.: Denoël, 2008).

Lerner, Daniel. *The Passing of Traditional Society: Modernizing the Middle East; With an Introduction by David Riesman* (New York/London: Free Press/Collier-Macmillan, 1965 [1958]).

Lethen, Helmut. *Cool Conduct: The Culture of Distance in Weimar Germany*, trans. Don Reneau (Berkeley: University of California Press, 2002).

Lett, Denise Potrzeba. *In Pursuit of Status: The Making of South Korea's "New" Urban Middle Class* (Cambridge, MA: Harvard University Asia Center/Harvard University Press, 1998).

Levi, Giovanni. "Frail Frontiers?" *Past and Present* (2019), Supplement 14: 37–49.

Lewis, Sinclair. *Babbitt* (Harmondsworth: Penguin, 1985 [1923]).

Lezotte, Christine L. "Have You Heard the One about the Woman Driver? Chicks, Muscle, Pickups, and the Reimagining of the Woman behind the Wheel," unpublished doctoral dissertation (Bowling Green, OH: Bowling Green State University, May 2015).

Li, Chunling. "Profile of China's Middle Class," in Chunling Li (ed.), *The Rising Middle Classes in China* (n.p.: Paths International/Social Sciences Academic Press [China], 2012), 87–125.

Lieb, Claudia. *Crash: Der Unfall der Moderne* (Bielefeld: Aisthesis, 2009).

Lifset, Robert D. "A New Understanding of the American Energy Crisis of the 1970," in "The Energy Crises of the 1970s: Anticipations and Reactions in the Industrialized World," special issue, *Historical Social Research/Historische Sozialforschung* 39, no. 4 (2014): 22–42.

Lin, Xiao, Peter Wells, and Benjamin Sovacool. "The Death of a Transport Regime? The Future of Electric Bicycles and Transportation Pathways for Sustainable Mobility in China," *Technology Forecasting & Social Change* 132 (2018): 255–67.

Lindegaard, Laura Bing. "The Discursive Accomplishment of Rationalities in the Automobility Regime," in Marcel Endres, Katharina Manderscheid, and Christophe Mincke (eds.), *The Mobilities Paradigm: Discourses and Ideologies* (London: Routledge, 2016), 68–87.

Link, Jürgen. "Maintaining Normality: On the Strategic Function of the Media in Wars of Extermination," *Cultural Critique* 19 (Fall 1991): 55–65.

*Lloyd, David W. *Battlefield Tourism: Pilgrimage and the Commemoration of the Great War in Britain, Australia and Canada, 1919–1939* (Oxford: Berg, 1998).

*Löfgren, Orvar. *On Holiday: A History of Vacationing* (Berkeley: University of California Press, 1999).

Logemann, Jan L. *Trams or Tailfins? Public and Private Prosperity in Postwar West Germany and the United States* (Chicago: University of Chicago Press, 2012).

Lone, Stewart. "Japan and the Age of Speed: Urban Life and the Automobile 1925–1930" (discussion paper No. IS/05/494, Suntory Centre, Suntory and Toyota International Centres for Economic and Related Disciplines, London School of Economics and Political Science, July 2005), 1–13.

*Louter, David. *Windshield Wilderness: Cars, Roads, and Nature in Washington's National Parks* (Seattle: University of Washington Press, 2006).

Lowerson, John. *Sport and the English Middle Classes 1870–1914* (Manchester: Manchester University Press, 1993).

*Luckin, Bill. "War on the Roads: Traffic Accidents and Social Tension in Britain, 1939–45," in Roger Cooter and Bill Luckin (eds.), *Accidents in History: Injuries, Fatalities and Social Relations* (Amsterdam: Rodopi, 1997), 234–54.

MacKintosh, Phillip Gordon. "A Bourgeois Geography of Domestic Bicycling: Using Public Space Responsibly in Toronto and Niagara-on-the-Lake, 1890–1900," *Journal of Historical Sociology* 20, no. 1/2 (March/June 2007): 126–57.

*Mackintosh, Philip Gordon, and Glen Norcliffe. "Men, Women and the Bicycle: Gender and the Social Geography of Cycling in the Late Nineteenth-Century," in: Dave Horton, Paul Rosen and Peter Cox (eds.), *Cycling and Society* (Aldershot/Burlington: Ashgate, 2011), 153–77.

*Macpherson, Heidi Slettedahl. *Women's Movement: Escape as Transgression in North American Feminist Fiction* (Amsterdam: Rodopi, 2000).

Madeley, Christopher. "Kaishinsa, DAT, Nissan and the British motor vehicle industry" (discussion paper No. IS/05/494, London School of Economics and Political Science, July 2005), 14–38.

Madhur, Srinivasa, Ganeshan Wignaraha, and Peter Darjes. *Roads for Asian Integration: Measuring ADB's Contribution to the Asian Highway Network* (n.p.: Asian Development Bank, November 2009) (ADB Working Paper Series on Regional Economic Integration, No. 37).

Madsen, Grant. *Sovereign Soldiers: How the U.S. Military Transformed the Global Economy after World War II* (Philadelphia: University of Pennsylvania Press, 2018) (American Business, Politics, and Society, eds. Andrew Wender Cohen et al.).

Mailer, Norman. "The White Negro: Superficial Reflections on the Hipster," in Norman Mailer, *Advertisements for Myself* (Cambridge, MA: Harvard University Press, 1992 [1959]), 337–58.

Manderscheid, Katharina. "The Movement Problem, the Car and Future Mobility Regimes: Automobility as Dispositif and Mode of Regulation," *Mobilities* 9, no. 4 (2014): 604–26.

Mangan, J. A. (ed.). *A Sport-Loving Society: Victorian and Edwardian Middle-Class England at Play* (Abingdon: Routledge, 2006).

Manning, Patrick. "Global History and Maritime History," *International Journal of Maritime History* 25, no. 1 (June 2013): 1–22.

——. *Navigating World History: Historians Create a Global Past* (New York: Palgrave Macmillan, 2003).

——. *Slavery, Colonialism and Economic Growth in Dahomey, 1640–1960* (Cambridge: Cambridge University Press, 1982).

*Mårald, Erland. "Methanol as Future Fuel: Efforts to Develop Alternative Fuels in Sweden after the Oil Crisis," *History and Technology* 26, no. 4 (December 2010): 335–57.

Marchis, Vittorio. "Not Cars Alone: An Italian Story," in *Motu Proprio: ACI's First Hundred Years, Heading Onward* (Rome: Automobile Club d'Italia, October 2005), 87–94.

Martin, George. "The Global Intensification of Motorization and Its Impacts on Urban Social Ecologies," in Jim Conley and Arlene Tigar McLaren (eds.), *Car Troubles: Critical Studies of Automobility and Auto-Mobility* (Farnham: Ashgate, 2009), 219–33.

Marx, Leo. *The Pilot and the Passenger: Essays on Literature, Technology, and Culture in the United States* (New York: Oxford University Press, 1988).

Matsuda, Matt K. *Pacific Worlds: A History of Seas, Peoples, and Cultures* (Cambridge: Cambridge University Press, 2012).

Mazauric, Catherine. *Mobilités d'Afrique en Europe: Récits et figures de l'aventure* (Paris: Karthala, 2012).

McAllister, Janet Elspeth. "Stuck Fast: A Critical Analysis of the 'New Mobility Paradigm,'" master's thesis (Auckland: University of Auckland, 2011).

*McCarthy, Tom. *Auto Mania: Cars, Consumers, and the Environment* (New Haven, CT: Yale University Press, 2007).

McCright, Aaron M., and Riley E. Dunlap. "Defeating Kyoto: The Conservative Movement's Impact on U.S. Climate Change Policy," *Social Problems* 50, no. 3 (August 2003): 348–73.

McDonald, Kate. "Imperial Mobility: Circulation as History in East Asia under Empire," *Transfers* 4, no. 3 (Winter 2014): 68–87.

McGaughey, William H. *American Automobile Album* (New York: E. P. Dutton, 1954).

McHugh, Kathleen. "Women in Traffic: L.A. Autobiography," *South Atlantic Quarterly* 97, no 2 (Spring 1998): 391–412.

*McIntyre, Stephen L. "The Failure of Fordism: Reform of the Automobile Repair Industry, 1913–1940," *Technology and Culture* 41, no. 2 (April 2000): 269–99.

*——. "'The Repair Man Will Gyp You': Mechanics, Managers, and Customers in the Automobile Repair Industry, 1896–1940," unpublished dissertation (Columbia: University of Missouri-Columbia, May 1995).

*McKenzie, Brian A. "Creating a Tourist's Paradise: The Marshall Plan and France, 1948 to 1952," *French Politics, Culture & Society* 21, no. 1 (Spring 2003): 35–54.

McKeown, Adam. "Global Migration, 1846–1940," *Journal of World History* 15, no. 2 (2004): 155–89.

*McShane, Clay. *Down the Asphalt Path: The Automobile and the American City* (New York: Columbia University Press, 1994) (Columbia History of Urban Life, ed. Kenneth T. Jackson).

*McShane, Clay, and Joel A. Tarr. *The Horse in the City: Living Machines in the Nineteenth Century* (Baltimore: Johns Hopkins University Press, 2007).

McShane, Mary, Masaki Koshi, and Olof Lundin. "Public Policy toward the Automobile: A Comparative Look at Japan and Sweden," *Transportation Research Part A* 18A, no. 2 (1984): 97–109.

Menoret, Pascal. *Joyriding in Riyadh: Oil, Urbanism, and Road Revolt* (New York: University of Cambridge Press, 2014) (Cambridge Middle East Studies, eds. Charles Tripp et al.).

*Merki, Christoph Maria. "The Birth of Motoring out of Sport: Car Racing as a Public Relations Strategy, 1894–1905," in Laurent Tissot and Béatrice Veyrassat (eds.),

Technological Trajectories, Markets, Institutions: Industrialized Countries, 19th–20th Centuries; From Context Dependency to Path Dependency (Bern: Peter Lang, 2002), 227–49.

* ——. *Der holprige Siegeszug des Automobils 1895–1930: Zur Motorisierung des Strassenverkehrs in Frankreich, Deutschland und der Schweiz* (Vienna: Böhlau, 2002).

*Merriman, Peter. *Driving Spaces; A Cultural-Historical Geography of England's M1 Motorway* (Malden, MA/Oxford/Carlton, Austr.: Blackwell, 2007) (The Royal Geographical Society [with the Institute of British Geographers] Book Series).

——. "Rethinking Mobile Methods," *Mobilities* 9, no. 2 (2013): 167–87.

Middell, Matthias and Katja Naumann. "Global History and the Spatial Turn: From the Impact of Area Studies to the Study of Critical Junctures of Globalization," *Journal of Global History* 5 (2010): 149–70.

*Miller, Shawn William. *An Environmental History of Latin America* (Cambridge: Cambridge University Press, 2007) (New Approaches to the Americas, ed. Stuart Schwartz).

* ——. *The Street is Ours: Community, the Car, and the Nature of Public Space in Rio de Janeiro* (Cambridge: Cambridge University Press, 2018).

Mills, Katherine Lawrie. "Remapping the Road Story: Contemporary Narratives of Autonomy and Mobility in American Literature, Film, and Television," unpublished dissertation (Los Angeles: University of Southern California, May 2000).

* ——. *The Road Story and the Rebel: Moving through Film, Fiction, and Television* (Carbondale: Southern Illinois University Press, 2006).

Mitchell, Timothy. *Carbon Democracy: Political Power in the Age of Oil* (London: Verso, 2013 [2011]).

Mitchell, William J., Christopher E. Borroni-Bird, and Lawrence D. Burns. *Reinventing the Automobile: Personal Urban Mobility for the 21st Century* (Cambridge, MA: MIT Press, 2010).

*Milor, Alice. "Whose Business is Road Safety? From a Fragmented to an Integrated Approach in France and Europe (1972–1998)," *Transfers* 9, no. 3 (Winter 2019): 41–60.

Mittal, Gaurav. "The State and the Production of Informalities in Urban Transport: *Vikrams* in Dehradun, India," *Geoforum* 136 (2022): 273–82.

*Mo, Yajun. *Touring China: A History of Travel Culture, 1912–1949* (Ithaca, NY: Cornell University Press, 2021).

*Mohun, Arwen P. *Risk: Negotiating Safety in American Society* (Baltimore: Johns Hopkins University Press, 2013).

*Mom, Gijs. *Atlantic Automobilism: Emergence and Persistence of the Car, 1895–1940* (New York: Berghahn, 2015).

* ——. "Building an Automobile System: PIARC and Road Safety (1908–1938)" (Proceedings of the PIARC conference, Paris, 17–23 September 2007).

——. "Civilized Adventure as a Remedy for Nervous Times: Early Automobilism and Fin de Siecle Culture," *History of Technology* 23 (2001): 157–90.

——. "Clashes of Cultures: Road vs. Rail in the North-Atlantic World during the Inter-War Coordination Crisis," in Christopher Kopper and Massimo Moraglio (eds.), *The Organization of Transport: A History of Users, Industry, and Public Policy* (London: Routledge, 2015), 8–31.

——. "Compétition et coexistence: La motorisation des transports terrestres et le lent processus de substitution de la traction équine," *Mouvement Social* 229 (October/December 2009): 13–39.

——. "Conceptualising Technical Change: Alternative Interaction in the Evolution of the Automobile," in Helmuth Trischler and Stefan Zeilinger (eds.), *Tackling Transport* (London: Science Museum, 2003), 9–46.

*——. "Constructing Multifunctional Networks: Road Building in the Netherlands, 1810–1980," in Gijs Mom and Laurent Tissot (eds.), *Road History: Planning, Building and Use* (Lausanne: Alphil, 2007), 33–62.

*——. "Costs, Technology and Culture—Propelling the Early Taxicab, 1900–25," *Journal of Transport History* 24, no. 2 (September 2003): 199–221.

——. "Decentering Highways: European National Road Network Planning from a Transnational Perspective," in Hans-Liudger Dienel and Hans-Ulrich Schiedt (eds.), *Die moderne Strasse: Planung, Bau und Verkehr vom 18. bis zum 20. Jahrhundert* (Frankfurt: Campus, 2010), 77–100.

——. "The Dual Nature of Technology: Automotive Ignition Systems and the Evolution of the Car," in Christian Huck and Stefan Bauernschmidt (eds.), *Travelling Goods, Travelling Moods: Varieties of Cultural Appropriation (1850–1950)* (Frankfurt: Campus, 2012), 189–207.

*——. *The Electric Vehicle: Technology and Expectations in the Automobile Age* (Baltimore: Johns Hopkins University Press, 2004).

*——. "The Emancipation of the Urban View: Dutch Spatial Planning in an International Context (1920–1950)," in Jos Arts, Ruud Filarski, Hans Jeekel, and Bert Toussaint (eds.), *Builders and Planners: A History of Land-Use and Infrastructure Planning in the Netherlands* (Delft: Eburon, 2016), 85–149.

*——. "Encapsulating Culture: European Car Travel 1900–1940," *Journal of Tourism History* 3, no. 3 (November 2011): 289–307.

*——. *The Evolution of Automotive Technology: A Handbook* (Warrendale, PA: SAE International, 2014).

——. "'The Future is a Shifting Panorama': The Role of Expectations in the History of Mobility," in Weert Canzler and Gert Schmidt (eds.), *Zukünfte des Automobils: Aussichten und Grenzen der autotechnischen Globalisierung* (Berlin: edition sigma, 2008), 31–58.

——. "Frozen History: Limitations and Possibilities of Quantitative Diffusion Studies," in Ruth Oldenziel and Adri de la Bruhèze (eds.), *Manufacturing Technology, Manufacturing Consumers: The Making of Dutch Consumer Society* (Amsterdam: aksant, 2008), 73–94.

*——. *Globalizing Automobilism: Exuberance and the Emergence of Layered Mobility, 1900–1980* (New York: Berghahn, 2020).

*——. "Inter-Artefactual Technology Transfer: Road Building Technology in the Netherlands and the Competition between Bricks, Macadam, Asphalt and Concrete," *History and Technology* 20, no. 1 (April 2004): 3–23.

——. "De moderne elektro-auto," in Gijs Mom and Vincent van der Vinne, *De elektro-auto: Een paard van Troje?* (Deventer: Kluwer Voertuigtechniek, 1995), 9–110.

——. "New Mobility Studies: Niche to Mainstream; History and Social Sciences in the Mobility Turn" (*Transfers*, forthcoming).

——. "'Our City is Ruined. Reason: Road Widening': A Review Essay on New Mobility Studies, Social Science, and History," *Transfers* 12, no. 3 (Winter 2022): 95–105.

*——. *Pacific Automobilism: Adventure, Status and the Carnival of Mobility, 1975–2015* (New York: Berghahn, 2022).

*——. "Struggle of the Systems: Freight Mobility from a Transatlantic Perspective, 1920–2000," in Gérard Duc, Olivier Perroux, Hans-Ulrich Schiedt, and François

Walter (eds.), *Histoire des transports et de la mobilité: Entre concurrence modale et coordination (de 1918 à nos jours)/Transport and Mobility History: Between Modal Competition and Coordination (from 1918 to the Present)* (Neuchâtel: Alphil, 2014), 177–95.

——. "Trending *Transfers*: A Decade of New Mobility Studies through the Lens of Transmodality, Transnationalism, and Transdisciplinarity," *Transfers* 10, no. 1 (Spring 2020): 2–21.

*Mom, Gijs, and Ruud Filarski. *Van transport naar mobiliteit: De mobiliteitsexplosie (1895–2005)* (Zutphen: Walburg Pers, 2008).

Mom, Gijs, and Diekus van der Wey. "Ethanolverdamping en de dieselmotor: Literatuuronderzoek naar en ontwerp van een ethanolverdamper t.b.v. het gecombineerd gebruik van dieselbrandstof en ethanolgas in een DAF-dieselmotor" [Ethanol evaporation and the diesel engine: Literature research on and design of an ethanol evaporator to combine diesel fuel and ethanol gas in a DAF diesel engine], Bachelor of Science thesis (Apeldoorn: HTS-Autotechniek, 7 June 1982).

Montagu of Beaulieu, Lord, and E. Wilson McComb. *Behind the Wheel: The Magic and Manners of Early Motoring* (New York: Paddington, 1977).

*Moraglio, Massimo. *Storia delle prime autostrade Italiane (1922–1943): Modernizzazione, affari e propaganda* (Turin: Trauben, 2007).

Morgan, David. "Class and Masculinity," in Michael S. Kimmel, Jeff Hearn, and R. W. Connell (eds.), *Handbook of Studies on Men & Masculinities* (Thousand Oaks, CA: Sage, 2005), 165–77.

Morris, Jeremy. *Everyday Post-Socialism: Working-Class Communities in the Russian Margins* (London: Palgrave Macmillan/Springer Nature, 2016).

Morris, Jeremy, and Abel Polese. "Introduction: Informality—Enduring Practices, Entwined Livelihoods," in Jeremy Morris and Abel Polese (eds.), *The Informal Post-Socialist Economy: Embedded Practices and Livelihoods* (London: Routledge, 2014) (Routledge Contemporary Russia and Eastern Europe Series), 1–18.

*Möser, Kurt. "Autobasteln: Modifying, Maintaining, and Repairing Private Cars in the GDR, 1970–1990," in Lewis H. Siegelbaum (ed.), *The Socialist Car: Automobility in the Eastern Bloc* (Ithaca, NY: Cornell University Press, 2011), 157–69.

*——. *Fahren und Fliegen in Frieden und Krieg: Kulturen individueller Mobilitätsmaschinen 1880–1930* (Heidelberg: Verlag Regionalkultur, 2009) (Technik + Arbeit, 13; Schriften des Technoseum, Landesmuseum für Technik und Arbeit in Mannheim).

*——. *Geschichte des Autos* (Frankfurt: Campus, 2002).

Muchadenyika, Davison. "Informal Transport, Politics and Power in Harare," in Daniel E. Agbiboa (ed.), *Transport, Transgression and Politics in African Cities: The Rhythm of Chaos* (Abingdon: Routledge, 2019) (Cities and Society, ed. Chris Pickvance), 46–77.

Mukhopadhyay, Aparajita. *Imperial Technology and 'Native' Agency: A Social History of Railways in Colonial India 1850–1920* (London: Routledge, 2018).

——. "Transport Users, and Historie(s) of Transport: 'A View from Below,'" *Journal of Transport History* (2023): 1–15.

Muleev, E. "Why Do Marshrutkas Exist in One City and Not in Others? Toward a Political Economy of Routes in Russian Urban Public Transportation," *Journal of Economic Sociology* 21, no. 2 (March 2020): 99–114.

——. "*Skotovoz, pripiski* and the 'Law of Labour Settlement': Transportation Planning and Management in the USSR," *Journal of Transport History* 45, no. 1 (2024): 62–82.

*Müller, Dorit. "Transfers between Media and Mobility: Automobilism, Early Cinema and Literature, 1900–1920," *Transfers* 1, no. 1 (Spring 2011): 53–75.

*——. *Gefährliche Fahrten: Das Automobil in Literatur und Film um 1900* (Würzburg: Königshausen & Neumann, 2004) (Epistemata, Würzburger wissenschaftliche Schriften, Reihe Literaturwissenschaft, 486).

Müller, Dorit, and Heike Weber. "'Traffic'—On the Historical Alignment of Media and Mobility," *Transfers* 3, no. 1 (Spring 2013): 65–74.

Müller, Martin. "In Search of the Global East: Thinking between North and South," *Geopolitics* 25, no. 3 (2020): 734–55.

*Mulley, Corinne. "The Background to Bus Regulation in the 1930 Road Traffic Act: Economic, Political and Personal Influences in the 1920s," *Journal of Transport History* 4, no. 2 (September 1983): 1–19.

*Mumford, Lewis. *The Highway and the City* (New York/Toronto: New American Library of World Literature/New American Library of Canada, 1964).

*Murphy, Dervla. *Full Tilt: Ireland to India with a Bicycle* (London: Eland, 2010 [1965]).

Mutongi, Kenda. *Matatu: A History of Popular Transportation in Nairobi* (Chicago: University of Chicago Press, 2017).

Nail, Thomas. "What is the Philosophy of Movement?" *Mobility Humanities* 1, no. 1 (January 2022): 6–22.

*Neiertz, Nicolas. *La coordination des transports en France: De 1918 à nos jours* (Paris: Ministère de l'Économie, des Finances et de l'Industrie, Comité pour l'histoire économique et financière, 1999).

*Neumaier, Christopher. *Dieselautos in Deutschland und den USA: Zum Verhältnis von Technologie, Konsum und Politik, 1949–2005* (Stuttgart: Franz Steiner, 2010) (Transatlantische historische Studien; Veröffentlichungen des Deutschen Historischen Instituts Washington, DC, eds. Hartmut Berghoff, Martin Klimke, Anke Ortlep, and Corinna R. Unger, 43).

——. "Eco-Friendly versus Cancer Causing: Perceptions of Diesel Cars in West Germany and the United States 1970–1990," *Technology and Culture* 55, no. 2 (April 2014): 429–60.

Newman, Peter W. G., and Jeffrey R. Kenworthy. *Cities and Automobile Dependence: A Sourcebook* (Aldershot: Avebury Technical, 1992 [1989]).

Nickles, Shelley. "More is Better: Mass Consumption, Gender, and Class Identity in Postwar America," *American Quarterly* 54, no. 4 (December 2002): 581–622.

Noel, Lance, Benjamin K. Sovacool, Johannes Kester, and Gerardo Zarazua de Rubens. "Conspicuous Diffusion: Theorizing How Status Drives Innovation in Electric Mobility," *Environmental Innovation and Social Transitions* 31 (2019): 154–69.

Norcliffe, Glen. *Critical Geographies of Cycling: History, Political Economy and Culture* (Farnham: Ashgate, 2015).

*——. *The Ride to Modernity: The Bicycle in Canada, 1869–1900* (Toronto: University of Toronto Press, 2001).

Norton, Peter David. *Fighting Traffic: The Dawn of the Motor Age in the American City* (Cambridge, MA: MIT Press, 2008).

Nye, Joseph S., Jr. *Soft Power: The Means to Success in World Politics* (New York: Public Affairs, 2004).

*Nwabughuogu, Anthony I. "The Role of Bicycle Transport in the Economic Development of Eastern Nigeria, 1930–45," *Journal of Transport History* 5, no. 1 (March 1984): 91–98.

O'Connell, Sean. *The Car and British Society: Class, Gender and Motoring, 1896–1939* (Manchester: Manchester University Press, 1998).

Offer, Avner. *The Challenge of Affluence: Self-Control and Well-Being in the United States and Britain since 1950* (Oxford: Oxford University Press, 2006).

*Oldenziel, Ruth, and Adri Albert de la Bruhèze. "Contested Spaces: Bicycle Lanes in Urban Europe, 1900–1995," *Transfers* 1, no. 2 (Summer 2011): 29–49.

Orsato, Renato J., Marc Dijk, René Kemp, and Masaru Yarime. "The Electrification of Automobility: The Bumpy Ride of Electric Vehicles toward Regime Transition," in Frank W. Geels, René Kemp, Geoff Dudley, and Glenn Lyons (eds.), *Automobility in Transition? A Socio-Technical Analysis of Sustainable Transport* (New York: Routledge, 2012), 205–28.

*Orselli, Jean. "Usages et usagers de la route: Pour une histoire de moyenne durée, 1860–2008," doctoral dissertation (Paris: Université de Paris I Panthéon Sorbonne, 2009).

*——. *Usages et usagers de la route: Requiem pour un million de morts: 1860–2010* (Paris: L'Harmattan, 2011).

Oteng-Ababio, Martin, and Ernest Agyemang. "The Okada War in Urban Ghana: A Polemic Issue or Policy Mismatch?" *African Studies Quarterly* 15, no. 4 (September 2015): 25–44.

——. "Virtue out of Necessity? Urbanisation, Urban Growth and Okada Services in Accra, Ghana," *Journal of Geography and Geology* 4, no. 1 (March 2012): 148–62.

Oviedo, Daniel, Caren Levy, and Julio D. Dávila. "Constructing Wellbeing, Deconstructing Urban (Im)Mobilities in Abuja, Nigeria," in Tanu Priya Uteng and Karen Lucas (eds.), *Urban Mobilities in the Global South* (London: Routledge, 2017), 173–94.

Owen, Wilfred. *Strategy for Mobility* (Washington, DC: Brookings Institution, 1964).

Pacey, Arnold, and Francesca Bray. *Technology in World Civilization: A Thousand-Year History; Revised and Expanded Edition* (Cambridge, MA: MIT Press, 2021 [1990]).

Packer, Jeremy. *Mobility without Mayhem: Safety, Cars, and Citizenship* (Durham, NC: Duke University Press, 2008).

Palat, Ravi Arvind. "Beyond Orientalism: Decolonizing Asian Studies," *Development and Society* 29, no. 2 (December 2000): 105–35.

——. "Miracles of the Day Before? The Great Asian Meltdown and the Changing World-Economy," *Development and Society* 28, no. 1 (July 1999): 1–47.

*Paolini, Federico. "Transport and the Environment in Italy (1950–2006)," *Economics and Policy of Energy and Environment* 2 (2012): 219–44.

*Papazafeiropoulou, Alexia Sodia. "Technology Users as Empirically-Trained Mechanics: Assembly and Decoration of Improvised Vehicles in Greece during and after World War II," *ICON: Journal of the International Committee for the History of Technology* 18 (2012): 157–78.

*Parchesky, Jennifer. "Women in the Driver's Seat: The Auto-Erotics of Early Women's Films," *Film History* 18, no. 2 (2006): 174–84.

Park, Woo Hyun. "Directional Monopoly in Colonial Railroad Construction: The Longitudinal Route Centrism of the Joseon Railroad" (Session "History of Korean Railroad: Geography, Geopolitics and Imaginative Geography," 2023 GMHC&T2M Annual Conference, Seoul, Konkuk University, 25–28 October 2023).

Parkin, Frank. "Class Stratification in Socialist Societies," *British Journal of Sociology* 20, no. 4 (December 1969): 355–74.

Parrinder, Patrick. "Biographical Note," in H. G. Wells, *The History of Mr. Polly* (London: Penguin, 2005 [1910]), vii–xii.

Passalacqua, Arnaud. "Mobility Techniques: Between Globalization and Local Solutions," in Guillaume Carnino, Liliane Hilaire-Pérez, and Jérôme Lamy (eds.), *Global*

History of Techniques (Nineteenth to Twenty-First Centuries) (Turnhout: Brepols, 2024), 351–62.

Patico, Jennifer. "To Be Happy in a Mercedes: Tropes of Value and Ambivalent Visions of Marketization," *American Ethnologist* 32, no. 3 (2005): 479–96.

Pattillo-McCoy, Mary. *Black Picket Fences: Privilege and Peril among the Black Middle Class* (Chicago: University of Chicago Press, 1999).

Peace, Adrian. "The Politics of Transporting," *Africa: Journal of the International African Institute* 58, no. 1 (1988): 14–28.

Pédale, Jehan de la. *Pédalons! Avec une préface de Richard O'Monroy (Pompon) et une lettre de Richard Lesclide* (Paris: Véloce-Sport, 1892).

Pelzer, Thorben. *Engineering Trouble: US–Chinese Experiences of Professional Discontent, 1905–1945* (Leiden: Brill, 2023) (China Studies, published for the Institute for Chinese Studies, University of Oxford, ed. Rana Mitter).

* Perkins, Anita. *Travel Texts and Moving Cultures: German Literature and the Mobilities Turn* (Oxford: Peter Lang, 2016).

Perkins, Greg, and Johann Peter Murmann. "What Does the Success of Tesla Mean for the Future Dynamics in the Global Automobile Sector?" *Management and Organization Review* 14, no. 3 (September 2018): 471–80.

Péteri, György. "Alternative Modernity? Everyday Practices of Elite Mobility in Communist Hungary, 1956–1980," in Lewis H. Siegelbaum, *The Socialist Car: Automobility in the Eastern Bloc* (Ithaca, NY: Cornell University Press, 2011), 47–68.

——. "Streetcars of Desire: Cars and Automobilism in Communist Hungary (1958–70)," *Social History* 34, no. 1 (February 2009): 1–28.

Petridou, Eleni, and Maria Moustaki. "Human Factors in the Causation of Road Traffic Crashes," *European Journal of Epidemiology* 16, no. 9 (2000): 819–26.

* Piglia, Melina. *Autos, rutas y turismo: El automóvil club argentino y el estado* (Buenos Aires: Siglo Veintiuno, 2014).

Piot, Charles. *Remotely Global: Village Modernity in West Africa* (Chicago: University of Chicago Press, 1999).

* Pirie, Gordon. "Automobile Organizations Driving Tourism in Pre-Independence Africa," *Journal of Tourism History* 5, no. 1 (2013): 73–91.

* Plowden, William. *The Motor Car and Politics in Britain* (Harmondsworth: Penguin, 1971).

* Ponsavady, Stephanie. *Cultural and Literary Representations of the Automobile in French Indochina: A Colonial Roadshow* (New York: Palgrave Macmillan, 2018) (Studies in Mobilities, Literature, and Culture).

Pooley, Colin, Jean Turnbull, and Mags Adams. *A Mobile Century? Changes in Everyday Mobility in Britain in the Twentieth Century* (Aldershot: Ashgate, 2005).

Pooley, Colin G., and Jean Turnbull. "Commuting, Transport and Urban Form: Manchester and Glasgow in the Mid-Twentieth Century," *Urban History* 27, no. 3 (2000): 360–83.

Popov, Vladimir. "The Culture of New Mobility in Russia: Networks and Flows Formation," *Mobilities* 7, no. 1 (February 2012): 151–69.

* Porter, Gina. "Reflections on a Century of Road Transport Developments in West Africa and Their (Gendered) Impacts on the Rural Poor," *EchoGéo* 20 (April–June 2012): 1–14.

Potter, Christopher Thomas. "An Exploration of Social and Cultural Aspects of Motorcycling during the Interwar Period," unpublished PhD dissertation (Newcastle: University of Northumbria, August 2007).

*Pozharliev, Lyubomir. *The Road to Socialism: Transport Infrastructure in Socialist Bulgaria and Yugoslavia (1945–1989)* (Göttingen: V&R Unipress, 2023).

Prakash, Gyan. "Writing Post-Orientalist Histories of the Third World: Perspectives from Indian Historiography," *Comparative Studies in Society and History* 32, no. 2 (April 1990): 383–408.

*Primeau, Ronald. "From Ma Joad to Elizabeth Berg: Women on the Road in America," *Midamerica: The Yearbook of the Society for the Study of Midwestern Literature* xxvi (1999): 138–46.

*——. *Romance of the Road: The Literature of the American Highway* (Bowling Green, OH: Bowling Green State University Popular Press, 1996).

Prins, Annemiek. "Unfixing the City: Rickshaw Mobilities, Modernities and Urban Change in Dhaka," unpublished dissertation (Aberdeen: University of Aberdeen, 2020).

Prussi, M., et al. "JEC Well-to-Wheels Report V5: Well-to-Wheels Analysis of Future Automotive Fuels and Powertrains in the European Context" (Luxemburg: Publications Office of the European Union, 2020).

Quilantang, Rae Francis C., and Kimberly Nicole Panis Quilantang. "'Para po! (Stop, Please)': The Cultural Significance of Jeepney Decorations in the Philippines in the Context of Modernisation," *Mobility Humanities* 3, no. 1 (January 2024): 133–48.

Rabenstein, Rüdiger. "The History of German Workers' Cycling Association, Solidarity," *Cycling History 11* (Proceedings, 11th International Cycling History Conference, 23–25 August 2000): 160–67.

*——. *Radsport und Gesellschaft: Ihre socialgeschichtlichen Zusammenhänge in der Zeit von 1867 bis 1914* (Hildesheim: Weidmann, 1991).

*Radkau, Joachim. "Auto-Lust: Zur Geschichte der Geschwindigkeit," in Tom Koenigs and Roland Schaeffer (eds.), *Fortschritt vom Auto? Umwelt und Verkehr in den 90er Jahren: Kongress des Umwelt Forum Frankfurt a.M. am 5./6. Oktober 1990* (Munich: Raben Verlag von Wittern, 1991), 113–30.

*——. "'Die Nervosität des Zeitalters': Die Erfindung von Technikbedürfnissen um die Jahrhundertwende," *Kultur & Technik* 3 (1994): 51–57.

*Rae, John B. *The American Automobile Industry* (Boston: Twayne, 1984).

*——. "The Evolution of the Motor Bus as a Transport Mode," in Martin T. Farris and Paul T. McElhiney (eds.), *Modern Transportation: Selected Readings* (Boston: Houghton Mifflin, 1973 [1967]), 387–400 (reprint from *High Speed Ground Transportation Journal* 5 [Summer 1971]).

*——. *The Road and the Car in American Life* (Cambridge, MA: MIT Press, 1971).

Ravnbøl, Camilla Ida, Trine Mygind Korsby, and Anja Simonsen. "Introduction Special Section Transnational Street Business: Migrants in the Informal Economy," *Migration and Society: Advances in Research* 6 (2023): 1–15.

Regett, Anika, Wolfgang Mauch, and Ulrich Wagner. "Carbon Footprint of Electric Vehicles—A Plea for More Objectivity" (Forschungsstelle für Energiewirtschaft, 2019).

*Reinecke, Siegfried. *Autosymbolik in Journalismus, Literatur und Film: Strukturalfunktionale Analysen vom Beginn der Motorisierung bis zur Gegenwart* (Bochum: Universitätsverlag Dr. N. Brockmeyer, 1992) (Bochumer Studien zur Publizistik- und Kommunikationswissenschaft, ed. Heinz-Dietrich Fischer, 70).

Rekhviashvili, Lela. "We Need to Talk about Political Society: Subaltern Resistances beyond Civil Society in Eastern Europe and Eurasia," *Central Asian Survey* 42, no. 2 (2023): 219–37.

Rekhviashvili, Lela, and Thilo Lang. "Chinese Investments as Part of Infrastructure-Led Development: Multi-Scalar Contestations around Georgia's Flagship Infrastructure Projects," *Eurasian Geography and Economics*. https://doi.org/10.1080/15387216.2024.2311712.

Rekhviashvili, Lela, and Wladimir Sgibnev. "Theorising Informality and Social Embeddedness for the Study of Informal Transport: Lessons from the *Marshrutka* Mobility Phenomenon," *Journal of Transport Geography* 88 (2020): 102386.

*Renner, Andreas. "Watching Foreign Neighbours: Russian and Soviet Travel Writing about Japan in the First Half of the Twentieth Century," *Journal of Tourism History* 3, no. 1 (April 2011): 39–56.

*Rimmer, Peter J. *Rikisha to Rapid Transit: Urban Public Transport Systems and Policy in Southeast Asia* (Sydney: Pergamon Press, 1986).

Rizzo, Matteo. "Being Taken for a Ride: Privatisation of the Dar es Salaam Transport System 1983–1998," *Journal of Modern African Studies* 40, no. 1 (2002): 133–57.

Robinson, Ronald E. "Conclusion: Railways and Informal Empire," in Clarence B. Davis and Kenneth E. Wilburn Jr. (with Ronald E. Robinson) (eds.), *Railway Imperialism* (New York: Greenwood Press, 1991) (Contributions in Comparative Colonial Studies, No. 26), 175–96.

*Rodriguez, Louis. "The Development of the Truck: A Constructivist History," unpublished PhD thesis (Bethlehem, PA: Lehigh University, 1997).

*Röhnert, Jan (ed.). *Die Metaphorik der Autobahn: Literatur, Kunst, Film und Architektur nach 1945* (Cologne: Böhlau, 2014).

Roosen, Jorg, Wim Marneffe, and Lode Vereeck. "A Review of Comparative Vehicle Cost Analysis," *Transport Reviews* 35, no. 6 (2015): 720–48.

*Rose, Mark H. *Interstate: Express Highway Politics 1941–1956* (Lawrence: Regents Press of Kansas, 1979).

*Rose, Mark H., Bruce Seely, and Paul F. Barrett. *The Best Transportation System in the World: Railroads, Trucks, Airlines, and American Public Policy in the Twentieth Century* (Columbus: Ohio State University Press, 2006).

Rostow, W. W. *The Stages of Economic Growth: A Non-Communist Manifesto* (Cambridge: Cambridge University Press, 1971 [1960]).

Rubinstein, David. "Cycling in the 1890s," *Victorian Studies* 21, no. 1 (Autumn 1977): 47–71.

*Ruoff, Jeffrey (ed.). *Virtual Voyages: Cinema and Travel* (Durham, NC: Duke University Press, 2006).

Rycroft, Simon. "Changing Lanes: Textuality off and on the Road," *Transactions of the Institute of British Geographers* 21 (1996): 425–28.

Rynning, Maja Karoline, Tanu Priya Uteng, and Karen Lucas. "Epilogue: Creating Planning Knowledge through Dialogues between Research and Practice," in Tanu Priya Uteng and Karen Lucas (eds.), *Urban Mobilities in the Global South* (London: Routledge, 2017), 215–22.

Sachsenmaier, Dominic. *Global Perspectives on Global History: Theories and Approaches in a Connected World* (Cambridge: Cambridge University Press, 2011).

Sagaris, Lake, and Anvita Arora. "Cycling for Social Justice in Democratizing Contexts: Rethinking 'Sustainable' Mobilities," in Tanu Priya Uteng and Karen Lucas (eds.), *Urban Mobilities in the Global South* (London: Routledge, 2017), 19–40.

Salazar, Noel B., and Taehee Kim. "An Interview with Noel B. Salazar," *Mobility Humanities* 3, no. 1 (January 2024): 158–68.

Salmon, Paul, Michael G. Lenné, Neville A. Stanton, Daniel P. Jenkins, and Guy H. Walker. "Managing Error on the Open Road: The Contribution of Human Error Models and Methods," *Safety Science* 48 (2010): 1225–35.

Salvatore, Ricardo D. "Early American Visions of a Hemispheric Market in South America," in Berndt Ostendorf (ed.), *Transnational America: The Fading of Borders in the Western Hemisphere* (Heidelberg: Universitätsverlag C. Winter, 2002), 45–64.

Sanina, Anna. "The *Marshrutka* as a Socio-Cultural Phenomenon of a Russian Megacity," *City, Culture and Society* 2 (2011): 211–18.

Sapiro, Gisèle. *The Sociology of Literature*, trans. Madeline Bedecarré and Ben Libman (Stanford, CA: Stanford University Press, 2023 [2014]).

Savin, Ivan, and Jeroen van den Berg. "Main Topics in EIST during Its First Decade: A Computational-Linguistic Analysis," *Environmental Innovation and Societal Transitions* 41 (2021): 10–17.

*Scharff, Virginia. *Taking the Wheel: Women and the Coming of the Motor Age* (New York: UNM Press, 1991).

Schatzki, Theodore R. *Social Practices: A Wittgensteinian Approach to Human Activity and the Social* (Cambridge: Cambridge University Press, 1996).

Schewel, Kerilyn. "Understanding Immobility: Moving beyond the Mobility Bias in Migration Studies," *International Migration Review* 54, no. 2 (2019): 328–55.

*Schiffer, Michael Brian (with Tamara C. Butts and Kimberly K. Grimm). *Taking Charge: The Electric Automobile in America* (Washington, DC.: Diane, 1994).

Schindler, Seth, Jessica DiCarlo, and Dinesh Paudel. "The New Cold War and the Rise of the 21st-Century Infrastructure State," *Transactions of the Institute of British Geographers* 47 (2022): 331–46.

Schivelbusch, Wolfgang. *Geschichte der Eisenbahnreise: Zur Industrialisierung von Raum und Zeit im 19. Jahrhundert* (Munich: Wagenbach, 1977).

Schmucki, Barbara. "Against 'the Eviction of the Pedestrian': The Pedestrians' Association and Walking Practices in Urban Britain after World War II," *Radical History Review* 114 (Fall 2012): 113–38.

*——. "'If I Walked on My Own at Night I Stuck to Well Lit Areas': Gendered Spaces and Urban Transport in 20th Century Britain," *Research in Transportation Economics* 34 (2012): 74–85.

Schödermeier, Josef W. "Die Entwicklung und Zusammensetzung des Personenkraftwagenbestandes in Westeuropa (Eine retrospektive Analyse)," dissertation (Cologne: Universität Köln, 1961).

Schöller-Schwedes, Oliver, and Stephan Rammler. *Mobile Cities: Dynamiken weltweiter Stadt- und Verkehrsentwicklung* (Berlin: LIT, 2008).

Schönhammer, Rainer. *Das Leiden am Beifahren: Frauen und Männer auf dem Sitz rechts* (Göttingen: Vandenhoeck & Ruprecht, 1995).

Schot, Johan. "Confronting the Second Deep Transition through the Historical Imagination," *Technology and Culture* 57, no. 2 (April 2016): 445–56.

Schriber, Mary Suzanne. *Writing Home: American Women Abroad 1830–1920* (Charlottesville: University Press of Virginia, 1997).

Schwanen, Tim. "Towards Decolonised Knowledge about Transport," *Palgrave Communications: Humanities/Social Sciences/Business* 4 (2018): 1–6.

Seelinger, Matthew J. "From the Jeep to the Humvee: U.S. Army Light Combat Vehicles, World War II to the Present," *On Point* 11, no. 3 (Winter 2005/6): 8–13.

*Seely, Bruce. *Building the American Highway System: Engineers as Policy Makers* (Philadelphia: Temple University Press, 1987) (Technology and Urban Growth, eds. Blaine A. Brownell et al.).

*Segreto, Luciano, Carles Manera, and Manfred Pohl (eds.). *Europe at the Seaside: The Economic History of Mass Tourism in the Mediterranean* (New York: Berghahn, 2009) (Published in association with Institute for Corporate Culture Affairs [ICCA], eds. Hubert Bonin et al.).

Seiler, Cotton. *Republic of Drivers: A Cultural History of Automobility in America* (Chicago: University of Chicago Press, 2008).

——. "'So That We as a Race Might Have Something Authentic to Travel By': African American Automobility and Cold-War Liberalism," *American Quarterly* 58, no. 4 (December 2006): 1091–1117.

*Sellers, Christopher C. *Crabgrass Crucible: Suburban Nature and the Rise of Environmentalism in Twentieth-Century America* (Chapel Hill: University of North Carolina Press, 2012).

*Semmens, Kristin. *Seeing Hitler's Germany: Tourism in the Third Reich* (New York: Palgrave Macmillan, 2005).

Şengül, Buket, and Hamid Mostofi. "Impacts of E-Micromobility on the Sustainability of Urban Transportation—A Systematic Review," *Applied Sciences* 11 (2021): 5851.

*Shaffer, Marguerite S. *See America First: Tourism and National Identity, 1880–1940* (Washington, DC: Smithsonian Institution Press, 2001).

*Shapiro, Judith. *Mao's War Against Nature: Politics and the Environment in Revolutionary China* (Cambridge: Cambridge University Press, 2001).

Sharma, Rameshwar Dayal, Sandeep Jain, and Kewal Singh. "Growth Rate of Motor Vehicles in India—Impact of Demographic and Economic Development," *Journal of Economic and Social Studies* 1, no. 2 (July 2011): 137–53.

*Sheller, Mimi. *Aluminum Dreams; The Making of Light Modernity* (Cambridge, MA/ London: MIT Press, 2014).

——. *Advanced Introduction to Mobilities* (Cheltenham: Edward Elgar, 2021) (Elgar Advanced Introductions).

——. "Pluralizing Mobilities Theory for Post-Carbon Futures and Social Justice," *Transfers* 13, no. 1/2 (Spring/Summer 2023): 162–77.

*——. "Theorising Mobility Justice," *Tempo Social: Revista de Sociologia da USP* 30, no. 2 (May 2018): 17–34.

Sheller, Mimi, and John Urry, "The New Mobilities Paradigm," *Environment and Planning A* 38 (2006): 207–26.

Shimokawa, Koichi. "Japan: The Late Starter Who Outpaced All Her Rivals," in Theo Barker (ed.), *The Economic and Social Effects of the Spread of Motor Vehicles: An International Centenary Tribute* (London: MacMillan, 1987), 214–35.

Shnayerson, Michael. *The Car that Could: The Inside Story of GM's Revolutionary Electric Vehicle* (New York: Random House, 1996).

Siegelbaum, Lewis H. "Cars, Cars and More Cars: The Faustian Bargain of the Brezhnev Era," in Lewis H. Siegelbaum (ed.), *Borders of Socialism: Private Spheres of Soviet Russia* (New York: Palgrave Macmillan, 2006), 83–103.

——. "The Impact of Motorization on Soviet Society after 1945," in Manfred Krieger, Ulrike Gutzmann, and Dirk Schlinkert (eds.), *Towards Mobility: Varieties of Automobilism in East and West* (Wolfsburg: Volkswagen AG, 2009) (Schriften zur Unternehmensgeschichte von Volkswagen, 3), 21–29.

——. "Introduction," in: Lewis H. Siegelbaum (ed.). *The Socialist Car; Automobility in the Eastern Bloc* (Ithaca/London: Cornell University Press, 2011) 1–13.

——. "On the Side: Car Culture in the USSR, 1960s–1980s," *Technology and Culture* 50, no. 1 (January 2009): 1–23.

*——. "Roadlessness and the 'Path to Communism': Building Roads and Highways in Stalinist Russia," *Journal of Transport History* 29, no. 2 (2008): 277–94.

*—— (ed.). *The Socialist Car: Automobility in the Eastern Bloc* (Ithaca, NY: Cornell University Press, 2011).

*——. "Soviet Car Rallies of the 1920s and 1930s and the Road to Socialism," *Slavic Review* 64, no. 2 (Summer 2005): 247–73.

*Siegelbaum, Lewis H., and Leslie Page Moch. *Broad is My Native Land: Repertoires and Regimes of Migration in Russia's Twentieth Century* (Ithaca, NY: Cornell University Press, 2014).

Sietchiping, Remy, Melissa Jane Permezel, and Caude Ngomsi. "Transport and Mobility in Sub-Saharan Cities: An Overview of Practices, Lessons and Options for Improvement," *Cities* 29 (2012): 183–89.

*Sievers, Immo, *AutoCars: Die Beziehungen zwischen der englischen und der deutschen Automobilindustrie vor dem Ersten Weltkrieg* (Frankfurt am Main: Lang, 1995) (Europäische Hochschulschriften, Series III, Vol. 640).

Simone, AbdouMaliq. "The Urbanity of Movement: Dynamic Frontiers in Contemporary Africa," *Journal of Planning Education and Research* 31, no. 4 (2011): 379–91.

Simpson, Clare S. "Capitalising on Curiosity: Women's Professional Cycle Racing in the Late-Nineteenth Century," in Dave Horton, Paul Rosen, and Peter Cox (eds.), *Cycling and Society* (Aldershot: Ashgate, 2011), 47–65.

*Sjöblom, Gustav. "The Political Economy of Railway and Road Transport in Britain and Germany, 1918–1933," unpublished dissertation (Cambridge: Darwin College, University of Cambridge, 16 April 2007).

*——. "The Shift from Railways to Roads: Four Propositions about the Interwar Period in Britain and Germany," *Jahrbuch für Wirtschaftsgeschichte* 1 (2007): 55–65.

Skjølsvold, Tomas Moe, and Marianne Ryghaug. "Temporal Echoes and Cross-Geography Policy Effects: Multiple Levels of Transition Governance and the Electric Vehicle Breakthrough," *Environmental Innovation and Social Transitions* 35 (2020): 232–40.

*Smethurst, Paul. *The Bicycle—Towards a Global History* (New York: Palgrave Macmillan, 2015).

Sorin, Gretchen. *Driving While Black: African American Travel and the Road to Civil Rights* (New York: Liveright, 2020).

Sōseki, Natsume. "The Diary of a Bicycle Rider," in *Spring Miscellany and London Essays*, trans. and introduced by Sammy I. Tsunematsu (Boston: Tuttle, 2002 [1909]).

——. *The Wayfarer: Kōjin*, trans. and with an introduction by Beongcheon Yu (New York: Perigree, 1982 [1967]).

Sovacool, Benjamin. "Expanding Carbon Removal to the Global South: Thematic Concerns on Systems, Justice, and Climate Governance," *Energy and Climate Change* 4 (2023): 100103.

Sovacool, Benjamin, Wokje Abrahamse, Long Zhang, and Jingzheng Ren. "Pleasure or Profit? Surveying the Purchasing Intentions of Potential Electric Vehicle Adopters in China," *Transportation Research Part A* 124 (2019): 69–81.

Sperling, Daniel, with contributions by Anne Brown, Robin Chase, Michael J. Dunne, Susan Pike, Steven E. Polzin, Susan Shaheen, Brian D. Taylor, Levi Tilleman, and Ellen van der Meer. *Three Revolutions: Steering Automated, Shared, and Electric Vehicles to a Better Future* (Washington, DC: Island Press, 2018).

*Spode, Hasso. "Der Aufstieg des Massentourismus im 20. Jahrhundert," in Heinz-Gerhard Haupt and Claudius Torp (eds.), *Die Konsumgesellschaft in Deutschland 1890–1990: Ein Handbuch* (Frankfurt: Campus, 2009), 114–28.

Stallybrass, Peter, and Allon White. *The Politics and Poetics of Transgression* (Ithaca, NY: Cornell University Press, 1986).

Steele, M. William. "The Making of a Bicycle Nation: Japan," *Transfers* 2, no. 2 (Summer 2012): 70–94.

——. "A Modern Infrastructure of Manchukuo: Where Did All the Cement Come from?" Paper presented at the East Asian Environmental History Conference, Kagawa University, Japan (22–25 October 2015).

——. "The Speedy Feet of the Nation: Bicycles and Everyday Mobility in Modern Japan," *Journal of Transport History* 31, no. 2 (December 2010): 182–209.

*Stefan, Adelina Oana. "Automobility and the Building of Tourism in Communist Romania c. 1960–89," in Colin Divall (ed.), *Cultural Histories of Sociabilities, Spaces and Mobilities* (London: Pickering & Chatto, 2015) (Studies for the International Society for Cultural History, eds. Chris Dixon and Birgitta Svensson), 179–91.

*Stephens, Jennie C. "The Dangers of Masculine Technological Optimism: Why Feminist, Antiracist Values Are Essential for Social Justice, Economic Justice, and Climate Justice," *Environmental Values* 33, no. 1 (2024): 58–70.

*Stout, Janis P. *The Journey Narrative in American Literature: Patterns and Departures* (Westport, CT: Greenwood Press, 1983).

*——. *Through the Window, Out the Door: Women's Narratives of Departure, from Austin and Cather to Tyler, Morrison and Didion* (Tuscaloosa: University of Alabama Press, 1998).

Strand, David. *Rickshaw Beijing: City People and Politics in the 1920s* (Berkeley: University of California Press, 1989).

Streim, Gregor. "Flucht nach vorn zurück: Heinrich Hauser—Portrait eines Schriftstellers zwischen Neuer Sachlichkeit und 'reaktionärem Modernismus,'" *Jahrbuch der Deutschen Schillergesellschaft* 43 (1999): 377–402.

Subrahmanyam, Sanjay. "On the Origins of Global History: Inaugural Lecture Delivered on Thursday 28 November 2013; transl Liz Libbrecht." Retrieved 12 February 2025 from https://books.openedition.org/cdf/4171.

Sundén, Jenny. "A Sense of Play: Affect, Emotion and Embodiment in *World of Warcraft*," in Marianne Liljeström and Susanna Paasonen (eds.), *Working with Affect in Feminist Readings: Disturbing Differences* (London: Routledge, 2010).

*Sutter, Paul S. *Driven Wild: How the Fight Against Automobiles Launched the Modern Wilderness Movement* (Seattle: University of Washington Press, 2002) (Weyerhaeuser Environmental Books).

Taalbi, Josef. "On the gendering of the early American electric car," *Environmental Innovation and Societal Transitions* 54 (2025) 100934.

Tarr, Joel A., and Clay McShane. "The Horse as an Urban Technology," *Journal of Urban Technology* 15, no. 1 (2008): 5–17.

*Tasker, Yvonne (ed.). *Action and Adventure Cinema* (London: Routledge, 2004).

Temenos, Cristina, et al. "Theorizing Mobility Transitions: An Interdisciplinary Conversation," *Transfers* 7, no. 1 (Spring 2017): 113–29.

*Tetzlaff, Stefan. "Revolution or Evolution? The Making of the Automobile Sector as a Key Industry in Mid-Twentieth Century India," in Arve Hansen and Kenneth Bo Nielsen (eds.), *Cars, Automobility and Development in Asia: Wheels of Change* (London: Routledge, 2017) (Routledge Studies in Transport, Environment and Development), 62–79.

Therborn, Göran. "Dreams and Nightmares of the World's Middle Classes," *New Left Review* 124 (July August 2020): 63–87.

Theweleit, Klaus. *Männerphantasien 1 + 2; Band 1: Frauen, Fluten, Körper, Geschichte; Band 2: Männerkörper—zur Psychoanalyse des weissen Terrors* (Munich: Piper, 2009 [1977/78]).

*Thompson, Gregory Lee. *The Passenger Train in the Motor Age: California's Rail and Bus Industries, 1910–1941* (Columbus: Ohio State University Press, 1993) (Historical Perspectives on Business Enterprise).

Thrift, Nigel. "Intensitäten des Fühlens: Für eine räumliche Politik des Affekts," in Helmuth Berking (ed.), *Die Macht des Lokalen in einer Welt ohne Grenzen* (Frankfurt: Campus, 2006), 216–251.

——. *Spatial Formations* (London: Sage, 2013 [1996]).

Thrift, Nigel, Paul Harrison, and Ben Anderson. "'The 27th Letter': An Interview with Nigel Thrift," in Ben Anderson and Paul Harrison (eds.), *Taking-Place: Non-Representational Theories and Geography* (Farnham: Ashgate, 2010), 183–98.

Thrift, Nigel, Adam Tickell, Steve Woolgar, and William H. Rupp (eds.). *Globalization in Practice* (Oxford: Oxford University Press, 2014).

*Tichi, Cecilia. *Shifting Gears: Technology, Literature, Culture in Modernist America* (Chapel Hill: University of North Carolina Press, 1987).

*Tissot, Laurent (ed.). *Development of a Tourist Industry in the 19th and 20th Centuries: International Perspectives* (Neuchâtel: Alphil, 2003).

*——. *Naissance d'une industrie touristique: Les Anglais et la Suisse au XIXe siècle* (Lausanne: Éditions Payot, 2000).

*——. "Naissance d'une Europe ferroviaire: La convention internationale de Berne (1890)," in Michèle Merger and Dominique Barjot (eds.), *Les entreprises et leurs réseaux: Hommes, capitaux, techniques et pouvoirs, XIXe et XXe siècles; Mélanges en l'honneur de François Caron* (Paris: Presses de l'Université de Paris-Sorbonne, 1998), 283–95.

Tiwari, Geetam. "Low Cost Means of Transport in Cities: The Critical Element in City Transport System in Low Income Countries," in X. Godard and I. Fatonzoun (eds.), *Urban Mobility for All/La Mobilité Urbaine pour Tous: Proceedings of the Tenth International CODATU Conference, Lomé, Togo, 12–15 November 2002* (Lisse: A. A. Balkema, 2002), 215–20.

*Tobin, Gary Allen. "The Bicycle Boom of the 1890s: The Development of Private Transportation and the Birth of the Modern Tourist," *Journal of Popular Culture* 7, no. 4 (1974): 838–49.

Toivanen, Anna-Leena. *Mobilities and Cosmopolitanisms in African and Afrodiasporic Literatures* (Leiden: Rodopi, 2021) (Textxet, Studies in Comparative Literature, eds. Theo D'haen et al.).

Tomba, Luigi. *The Government Next Door: Neighborhood Politics in Urban China* (Ithaca, NY: Cornell University Press, 2014).

*Uekötter, Frank. *The Greenest Nation? A New History of German Environmentalism* (Cambridge, MA: MIT Press, 2014).

*——. "The Merits of the Precautionary Principle: Controlling Automobile Exhausts in Germany and the United States before 1945," in E. Melanie DuPuis (ed.), *Smoke and Mirrors: The Politics and Culture of Air Pollution* (New York: New York University Press, 2004), 119–53.

Urry, John. "Globale Komplexitäten," in Helmuth Berking (ed.), *Die Macht des Lokalen in einer Welt ohne Grenzen* (Frankfurt: Campus, 2006), 87–102.

——. "The 'System' of Automobility," *Theory, Culture and Society* 21, no. 4/5 (2004): 25–39.

——. *The Tourist Gaze: Leisure and Travel in Contemporary Societies* (London: Sage, 1990).

——. *What is the Future?* (Cambridge: Polity Press, 2016).

Uteng, Tanu Priya, and Karen Lucas. "The Trajectories of Urban Mobilities in the Global South: An Introduction," in Tanu Priya Uteng and Karen Lucas (eds.), *Urban Mobilities in the Global South* (London: Routledge, 2017), 1–18.

Vahl, H. G., and J. Giskes. *Traffic Calming through Integrated Urban Planning* (Paris: Amarcande, April 1990 [1988]).

Van den Boomen, Tijs. "Op naar de leesbare straat" [Towards the readable street], *NRC Handelsblad* (2 June 2001).

Van der Straeten, Jonas. "Sustainability's 'Other': Coming to Terms with the Electric Rickshaw in Bangladesh," *Historical Social Research* 47, no. 4 (2022): 139–67.

Varma, Pavan K. *The Great Indian Middle Class* (New Delhi: Penguin, 2007 [1998]).

Vasconcellos, Eduardo A. "Reassessing Traffic Accidents in Developing Countries," *Transport Policy* 2, no. 4 (1996): 263–69.

*——. *Urban Transport, Environment and Equity: The Case for Developing Countries* (London: Earthscan, 2001).

Verdery, Katherine. *What Was Socialism, and What Comes Next?* (Princeton, NJ: Princeton University Press, 1996).

Verstraete, Ginette. *Tracking Europe: Mobility, Diaspora, and the Politics of Location* (Durham, NC: Duke University Press, 2010).

Vielma-Cabruja, José Ignacio. "Transfers, Influences and Experiences: Notes on the Development of Expressways in Caracas (1947–1973) and Its Actual Expression" (paper presented at the workshop "Transport and Mobility Studies in Latin America," School of Architecture and Urban Studies of the University of Chile, Santiago de Chile, 8–9 July 2013).

*Vinsel, Lee Jared. "Federal Regulation Management of the Automobile in the United States," dissertation (Pittsburgh, PA: Carnegie Mellon University, May 2011).

Vivi, Matteo, and Anne-Mette Hermans. "'Zero Emission, Zero Compromises': An Intersectional, Qualitative Exploration of Masculinities in Tesla's Consumer Stories," *Men and Masculinities* 25, no. 4 (2022): 622–44.

Vogel, Ezra F. *Japan's New Middle Class: The Salary Man and His Family in a Tokyo Suburb* (Berkeley: University of California Press, 1971).

Vögele, Jörg, Stefanie Knöll, and Thorsten Noack (eds.). *Epidemics and Pandemics in Historical Perspective/Epidemien und Pandemien in historischer Perspektive* (Herbolzheim: Centaurus Press, 2015), 177–89.

*Volti, Rudi. *Cars and Culture: The Life Story of a Technology* (Westport, CT: Greenwood Press, 2004) (Greenwood Technographies).

*——. "Reducing Automobile Emissions in Southern California: The Dance of Public Policies and Technological Fixes," in Arthur Molella and Joyce Bedi, *Inventing for the Environment* (Cambridge, MA: MIT Press in association with the Lemelson Center, Smithsonian Institution, 2003) (Lemelson Center Studies in Invention and Innovation, eds. Arthur Molella and Joyce Bedi), 277–88.

*Von Saldern, Adelheid. "Cultural Conflicts, Popular Mass Culture, and the Question of Nazi Success: The Eilenriede Motorcycle Races, 1924–39," *German Studies Review* 15 (1992): 317–38.

Vuchic, Vukan R. *Transportation for Livable Cities* (New Brunswick, NJ: Center for Urban Policy Research, Rutgers, State University of New Jersey, 1999).

Vukovich, Daniel Frederick. "Sinological-Orientalism: The Production of the West's Post-Mao China," dissertation (Champaign: University of Illinois at Urbana-Champaign, 2005).

*Wakefield, Ernest Henry. *History of the Electric Automobile: Battery-Only Powered Cars* (Warrendale, PA: Society of Automotive Engineers, 1994).

*——. *History of the Electric Automobile: Hybrid Electric Vehicles* (Warrendale, PA: Society of Automotive Engineers, 1998).

Wallerstein, Immanuel. *World-Systems Analysis: An Introduction* (Durham, NC: Duke University Press, 2004).

*Walsh, Margaret. "Gender and American Mobility: Cars, Women and the Issue of Equality," in Colin Divall (ed.), *Cultural Histories of Sociabilities, Spaces and Mobilities* (London: Pickering & Chatto, 2015) (Studies for the International Society for Cultural History, eds. Chris Dixon and Birgitta Svensson), 29–37.

——. "Gendering Mobility: Women, Work and Automobility in the United States," *History* 93, no. 3 (July 2008): 376–95.

*——. *Making Connections: The Long-Distance Bus Industry in the USA* (Aldershot: Ashgate, 2000).

*Walton, John K. *The British Seaside: Holidays and Resorts in the Twentieth Century* (Manchester: Manchester University Press, 2000).

*——. (ed.). *Histories of Tourism: Representation, Identity and Conflict* (Clevedon: Channel View, 2005) (Tourism and Cultural Change, eds. Mike Robinson and Alison Phipps, 6).

Wappelhorst, Sandra. "The End of the Road? An Overview of Combustion-Engine Car Phase-Out Announcements across Europe" (Briefing, International Council on Clean Transportation, ICCT, 10 May 2020) (https://theicct.org/publication/the-end-of-the-road-an-overview-of-combustion-engine-car-phase-out-announcements-across-europe/, last consulted 13 February 2025).

*Warren, James Francis. *Rickshaw Coolie: A People's History of Singapore 1880–1940* (Singapore: Singapore University Press, 2003 [1986]).

Watson, Brian T. *Headed into the Abyss: The Story of Our Time, and the Future We'll Face* (Swampscott, MA: Anvilside Press, 2019).

Weber, Donald. *De blijde intrede van de automobiel in België 1895–1940* (Ghent: Academia Press in collaboration with AMSAB-ISG, 2010).

Wei, Hui [Zhou Weihui]. *Shanghai Baby*, trans. Bruce Humes (London: Constable, 2002 [1999]).

Weingroff, Richard F. "The Man Who Loved Roads," *Public Roads* 65, no. 6 (May/June 2002): 37–46.

*Wells, Christopher W. *Car Country: An Environmental History*, Foreword by William Cronon (Seattle: University of Washington Press, 2012).

——. "The Road to the Model T: Culture, Road Conditions, and Innovation at the Dawn of the American Motor Age," *Technology and Culture* 48, no. 3 (July 2007): 497–523.

Wells, H. G. *The History of Mr Polly*, ed. Simon J. James, with an Introduction by John Sutherland and Notes by John Sutherland and Simon J. James (London: Penguin, 2005 [1910]).

——. *The Wheels of Chance: A Holiday Adventure* (London/New York: J. M. Dent/Macmillan, 1896).

Wells, Peter, and Dimitrios Xenias. "From 'Freedom of the Open Road' to 'Cocooning': Understanding Resistance to Change in Personal Private Automobility," *Environmental Innovation and Societal Transitions* 16 (2015): 106–19.

Wenzlhuemer, Roland. *Globalgeschichte schreiben: Eine Einführung in 6 Episoden* (Konstanz: UVK Verlagsgesellschaft, 2017).

*Wesseling, Joeri. "Strategies of Car Manufacturers in Sustainable Transitions," dissertation (Utrecht: University of Utrecht, 2015).

*Wetmore, Jameson M. "Engineering with Uncertainty: Monitoring Air Bag Performance," *Science and Engineering Ethics* 14 (2008): 201–18.

Wetterauer, Andrea. *Lust an der Distanz: Die Kunst der Autoreise in der "Frankfurter Zeitung"* (Tübingen: Tübinger Verein für Volkskunde, 2007).

Wierzbicka, Anna. *Imprisoned in English: The Hazards of English as a Default Language* (Oxford: Oxford University Press, 2014).

Wieser, Matthias. *Das Netzwerk von Bruno Latour: Die Akteur-Netzwerk-Theorie zwischen Science & Technology Studies and poststrukturalistischer Soziologie* (Bielefeld: transcript, 2012).

Wightman, David. *Towards Economic Cooperation in Asia: The United Nations Economic Commission for Asia and the Far East* (New Haven, CT: Yale University Press, 1963).

*Wilkins, Mira, and Frank Ernest Hill. *American Business Abroad: Ford on Six Continents* (Detroit: Wayne State University Press, 1964).

Williamson, Thomas. "The Fluid State: Malaysia's National Expressway," *Space & Culture* 6, no. 2 (May 2003): 110–31.

Wilson, Fiona. "The Conflict between Indigenous and Immigrant Commercial Systems in the Peruvian Central Sierra, 1900–1940," in Rory Miller (ed.), *Region and Class in Modern Peruvian History* (Liverpool: University of Liverpool, Institute of Latin American Studies, 1987), 125–61.

Withers, Jeremy. *The War of the Wheels: H.G. Wells and the Bicycle* (Syracuse, NY: Syracuse University Press, 2017).

Wittrock, Björn. "Modernity: One, None, or Many? European Origins and Modernity as a Global Condition," *Daedalus* 129, no. 1 (Winter 2000): 31–60.

Wohl, Robert. *The Generation of 1914* (Cambridge, MA: Harvard University Press, 1979).

*Wolfe, Joel. *Autos and Progress: The Brazilian Search for Modernity* (Oxford: Oxford University Press, 2010).

Wolff, Pauline. "From the Urban Planning Discourse to a Circulation *Dispositif*: An Epistemological Approach to the Mobility Turn," in Marcel Endres, Katharina Manderscheid, and Christophe Mincke (eds.), *The Mobilities Paradigm: Discourses and Ideologies* (London: Routledge, 2016), 202–23.

*Wrobel, David M., and Patrick T. Long (eds.). *Seeing and Being Seen: Tourism in the American West* (Lawrence: University Press of Kansas, 2001).

*Xu, Tao. "Making a Living: Bicycle-Related Professions in Shanghai, 1897–1949," *Transfers* 3, no. 3 (Winter 2013): 6–26.

Yosano, Akiko. *Travels in Manchuria and Mongolia; A Feminist Poet from Japan Encounters Prewar China; Translated by Joshua A. Fogel* (New York: Columbia University Press, 2001).

*Zeller, Thomas. *Strasse, Bahn, Panorama: Verkehrswege und Landschaftsveränderung in Deutschland von 1930 bis 1990* (Frankfurt: Campus, 2002).

Zezina, Maria R. "The Introduction of Motor Vehicles on a Mass Scale in the USSR," in Manfred Grieger, Ulrike Gutzmann, and Dirk Schlinkert (ed.), *Towards Mobility: Varieties of Automobilism in East and West* (Wolfsburg: Volkswagen AG, 2009), 43–53.

Zhan, Lingjing, and Yanqun He. "Understanding Luxury Consumption in China: Consumer Perceptions of Best-Known Brands," *Journal of Business Research* 65 (2012): 1452–60.

Zhang, Jun. *Driving toward Modernity: Cars and the Lives of the Middle Class in Contemporary China* (Ithaca, NY: Cornell University Press, 2019).

——. "A History of Bicycle Mobility in Urban China: Shared Bikes, Technology, and Infrastructures (Part 2 of 2)" (Max Planck Institute for Social Anthropology, MoLab Inventory of Mobilities and Socioeconomic Changes, April 2022).

Žižek, Slavoj. *Violence: Six Sideways Reflections* (New York: Picador, 2008).

Zou, Xin, Hai L. Vu, and Helai Huang. "Fifty Years of Accident Analysis & Prevention: A Bibliometric and Scientometric Overview," *Accident Analysis & Prevention* 144 (2020): 1–26.

*Zuelow, Eric G. E. (ed.). *Touring beyond the Nation: A Transnational Approach to European Tourism History* (Farnham: Ashgate, 2011).

Zuev, Dennis, David Tyfield, and John Urry. "Where is the Politics? E-Bike Mobility in Urban China and Civilizational Government," *Environmental Innovation and Societal Transitions* 30 (2019): 19–32.

*Zurbrugg, Nicholas. "'Oh What a Feeling!'—The Literatures of the Car," in David Thoms, Len Holden, and Tim Claydon (eds.), *The Motor Car and Popular Culture in the 20th Century* (Aldershot: Ashgate, 1998), 9–27.

*Zytnicki, Colette. "'Fair l'Algérie agréable': Tourisme et colonisation en Algérie des années 1870 à 1962," *Le Mouvement Social* 242 (January–March 2013): 97–114.

Index

Page numbers in italics indicate a definition or crucial treatment of the term. Only those contemporary scholars have been included whose names appear in the text more than in passing.

www.ingramcontent.com/pod-product-compliance
Lightning Source LLC
Chambersburg PA
CBHW060633180825
31190CB00004B/38

9781836950912